GUIDE TO EXTENSION TRAINING

FAO Training Series *No. 11*

Guide to extension training

P. Oakley and C. Garforth

**Agriculture Extension and Rural Development Centre,
School of Education, University of Reading, UK**

FOOD AND AGRICULTURE ORGANIZATION OF THE UNITED NATIONS
Rome, 1985

The designations employed and the presentation of material in this publication do not imply the expression of any opinion whatsoever by the Food and Agriculture Organization of the United Nations concerning the legal status of any country, territory, city or area or of its authorities, or concerning the delimitation of its frontiers or boundaries. The views expressed are those of the authors.

P-67
ISBN 92-5-101453-1

Printed in Italy

Foreword

This *Guide to Extension Training* is a new edition of a text which was originally written by D.J. Bradfield in 1966 and later revised in 1969. In this new edition we have largely kept to the basic structure and broad content outline of Bradfield's 1969 revision, except for a completely new first chapter. We have, however, considerably reorganized the material and rewritten it entirely.

Bradfield's text was based almost wholly on extension experience in Malawi and drew its examples and approaches from that country. In this new edition we have drawn upon our joint experiences with extension in the three principal continental regions of the developing world — Asia, Africa and Latin America — and have used material from these regions in the text. Since the mid-1960s there have been a number of changes in the conception and practice of extension, and we have included such changes in the text, and generally brought it up to date.

The purposes of this guide are several. First, it is intended to be a text for those involved in the pre-service and in-service training of extension personnel. Second, we hope it can be used directly by extension agents in the field as a resource text in support of their extension activities. The text is a guide, and we have tried to lay out the material in an appropriate way. We hope that the style of the text will be useful for an extension agent who needs to understand the basic aspects of a particular extension issue.

The guide is directed toward extension agents in general. Of these, and given the importance of agriculture in rural areas, agricultural extension agents will be the greater number. The principles and methods of extension examined in this text are also relevant to those who work in extension in fields other than agriculture, such as home economists, community development workers or health workers.

This guide is written within the context of rural development and agricultural systems to be found in what we refer to as the developing world. We have drawn our material from extension practice in countries in Asia, Africa and Latin America. Although the principles of extension are applicable in any context, the analysis and discussion in this guide are in the context of the above three continental regions.

In writing this guide we have had to make decisions about the use of certain terms and in order to avoid misunderstanding we feel we ought to point these out.

When referring to the extension agent in the text we refer to *he* or *his*. This use of the masculine is not a lack of respect for female extension agents. However, it is our view that there are more male extension agents in different parts of the world and to use *he/she* or *his/her* jointly throughout the text would have been cumbersome. When we use *he* or *his,* therefore, we are referring to extension agents in general.

We use the term *farmer* throughout the text to refer to the rural people with whom extension agents work. We realize that this is a very general term, and that we cannot talk of the farmers as a whole when we refer to the different groups of rural people with whom extension works. We discuss this issue in detail in Chapter 3. Our use of the general term, farmer, is to indicate the extension client and is not intended to suggest that all rural people can be placed in the same category.

We would like to acknowledge the usefulness of the original text of D.J. Bradfield, which has served as a basic structure for this new edition of the guide. A word of thanks also to Christopher White for preparing the illustrations for the text so competently and under pressure of time.

Our thanks also to Diana McDowell, Lois Pegg and Jane Thompson for efficiently and cheerfully preparing and typing the text, and to Bridget Dillon for proof-reading. Finally, our appreciation to FAO for the opportunity to bring our different experiences together in the preparation of this guide. We certainly hope that it will prove useful to the many thousands of extension agents who work with millions of farmers throughout the world to increase food production, promote rural development and improve their standard of living.

Peter Oakley and Christopher Garforth
Reading, United Kingdom
October 1983

Contents

	Page
Foreword	**v**
1. The framework of development	**1**
The concept of development	1
Agricultural and rural development	2
Principles of rural development programmes	7
The importance of extension	8
2. Understanding extension	**9**
The concept of extension	9
Principles of extension	13
Extension and education	16
Types of extension	20
3. Social and cultural factors in extension	**23**
Social structure	23
Culture	29
Social and cultural change	33
Social and cultural barriers to agricultural change	37
4. Extension and communication	**41**
Communication	41
Mass media in extension	45
Audio-visual aids in extension	60
5. Extension methods	**67**
Individual methods of extension	68
Group methods of extension	75
Types of group extension methods	78
6. The extension agent	**91**
The role of the agent	92
Knowledge and personal skills	94
Public speaking	97
Report writing	100
The use of local leaders	101

7. The planning and evaluation of extension programmes **105**
Stages in programme planning 108
Evaluating extension programmes 114

8. Extension and special target groups **121**
Extension and rural women 121
Extension and rural youth 124
Extension and the landless 128

Bibliography **131**

Case-studies **133**

Index **141**

1. The framework of development

The concept of development

All rural extension work takes place within a process of development, and cannot be considered as an isolated activity. Extension programmes and projects and extension agents are part of the development of rural societies. It is, therefore, important to understand the term *development*, and to see how its interpretation can affect the course of rural extension work.

The term development does not refer to one single phenomenon or activity nor does it mean a general process of social change. All societies, rural and urban, are changing all the time. This change affects, for example, the society's norms and values, its institutions, its methods of production, the attitudes of its people and the way in which it distributes its resources. A rural society's people, customs and practices are never static but are continually evolving into new and different forms. There are different theories which seek to explain this process of social change (as evolution, as cultural adaptation or even as the resolution of conflicting interests) and examples of each explanation can be found in different parts of the world.

Development is more closely associated with some form of action or intervention to influence the entire process of social change. It is a dynamic concept which suggests a change in, or a movement away from, a previous situation. All societies are changing, and rural extension attempts to develop certain aspects of society in order to influence the nature and speed of the change. In the past few decades, different nations have been studied and their level of development has been determined; this has given rise to the use of terms such as *developed* as opposed to *developing* nations. In other words, it is assumed that some nations have advanced or changed more than others, and indeed these nations are often used as the model for other, developing, nations to follow.

This process of development can take different forms and have a variety of objectives. The following statements illustrate this:

- Development involves the introduction of new ideas into a social system in order to produce higher per caput incomes and levels of living through modern production methods and improved social organization.

• Development implies a total transformation of a traditional or pre-modern society into types of technology and associated social organization that characterize the advanced stable nations of the Western world.
• Development is building up the people so that they can build a future for themselves. Development is an experience of freedom in deciding what people choose to do. To decide to do something brings dignity and self-respect. Development efforts therefore start with the people's potential and proceed to their enhancement and growth.

Much has been written about the process of development, and the approaches which developing nations should adopt in order to develop. Reviewing this literature it can be concluded that a process of development should contain three main elements.

Economic. The development of the economic or productive base of any society, which will produce the goods and materials required for life.
Social. The provision of a range of social amenities and services (i.e., health, education, welfare) which care for the non-productive needs of a society.
Human. The development of the people themselves, both individually and communally, to realize their full potential, to use their skills and talents, and to play a constructive part in shaping their own society.

Development has to do with the above three elements. It should not concentrate upon one to the exclusion of the others. The economic base of any society is critical, for it must produce the resources required for livelihood. But we must also think of people and ensure their active participation in the process of development.

Agricultural and rural development

This guide is primarily concerned with rural extension and with the livelihoods of farmers and their families. The concept of rural development must therefore be considered with particular reference to agriculture, since agriculture is the basis of the livelihood of most rural families. In the past two decades there has been increasing emphasis on rural development programmes and projects, and recognition that the development of rural areas is just as important as the building up of urban, industrial com-

Rural development in an area of marginal rainfall in northern Ethiopia

plexes. Development must have two legs: urban industrialization and rural improvement.

There are very strong reasons why resources should now be put into rural development. More than half the people of the world and the vast majority of the people in developing countries (Asia, Africa and Latin America) live in rural areas and gain part or all of their livelihoods from some form of agriculture. Most of these people are also still very poor and dependent on agricultural practices that have benefited little from modern technology. They live in isolated and often inhospitable places, with little access to the resources they need to improve their agriculture. Many lead their lives barely at subsistence level. Solely in terms of numbers of people, there is a very strong case for giving high priority to rural development.

It can also be argued that agriculture is a vital part of the economy of any country and that its development is critical to the development of the country's economy as a whole. This relationship can perhaps be best understood by studying the following diagram (see p. 4).

Agriculture's important role is one of production, both of food for the rural and the urban population and of cash crops for the export market, to earn foreign currency. In this process demand is stimulated for other products and services, and employment opportunities emerge to

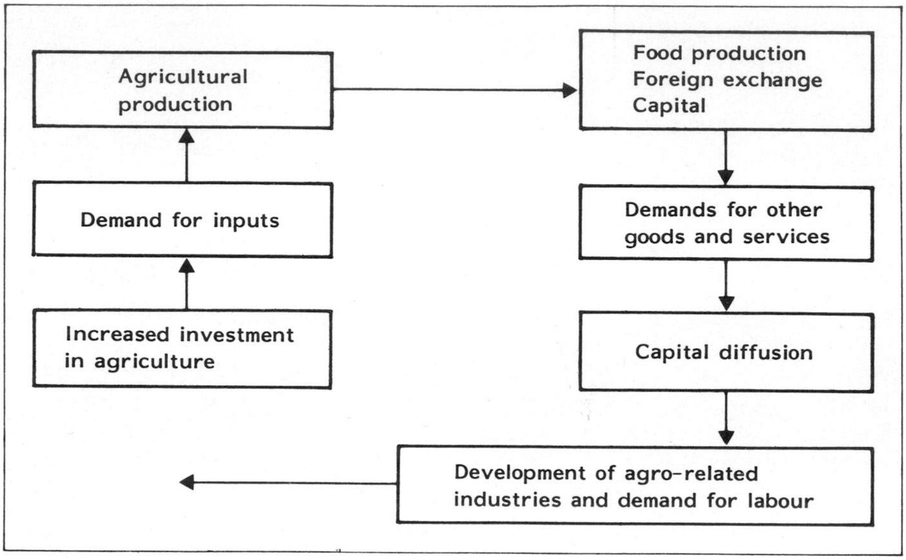

absorb the society's work-force. As the cycle develops, the increasing agricultural production causes an increasing demand for inputs, which ensure the resources required to maintain the agricultural production. Land is a basic resource for most countries and the exploitation of that resource in the interest of its citizens is one of a country's main responsibilities.

This concern to improve a country's agricultural base, and thus the livelihood of the majority of its inhabitants, is usually expressed in terms of programmes and projects of rural development. However, while agriculture is rightly the most important objective in the development of rural areas, rural development should also embrace the non-agricultural aspects of rural life. There are many definitions and statements on rural development that attempt to describe succinctly what it is trying to achieve. Perhaps the one used in conjunction with the UN-sponsored Second Development Decade in the 1970s best illustrates the broad nature of rural development:

- The Second Development Decade equates rural development with the far-reaching transformation of the social and economic structures, institutions, relationships and processes in any rural area. It conceives the goals of rural development not simply as agricultural and economic growth in the narrow sense but as balanced social and economic development.

Rural development is a process integrated with economic and social objectives, which must seek to transform rural society and provide a better and more secure livelihood for rural people. Rural development, therefore, is a process of analysis, problem identification and the proposal of relevant solutions. This process is usually encompassed within a programme or a project which seeks to tackle the problem identified.

However, as can be seen from the above statement, the problems that rural development programmes attempt to solve are not only agricultural; such programmes must also tackle the social or institutional problems found in rural areas. Indeed, if the kinds of problems which rural development programmes confront are considered in very broad terms, they may perhaps be divided into two.

Physical. These are problems which relate to the physical environment of a particular rural area, e.g., lack of water, poor infrastructure, lack of health facilities, or soil erosion. Rural development programmes can study the nature and extent of the problem and propose a course of action.

Non-physical. Not all the problems which farmers face are physical in nature. Some problems are more related to the social and political conditions of the region in which the farmers live, e.g., limited access to land, no contact with government services, or dependence upon a bigger farmer. These problems are also very real even though they exist below the surface.

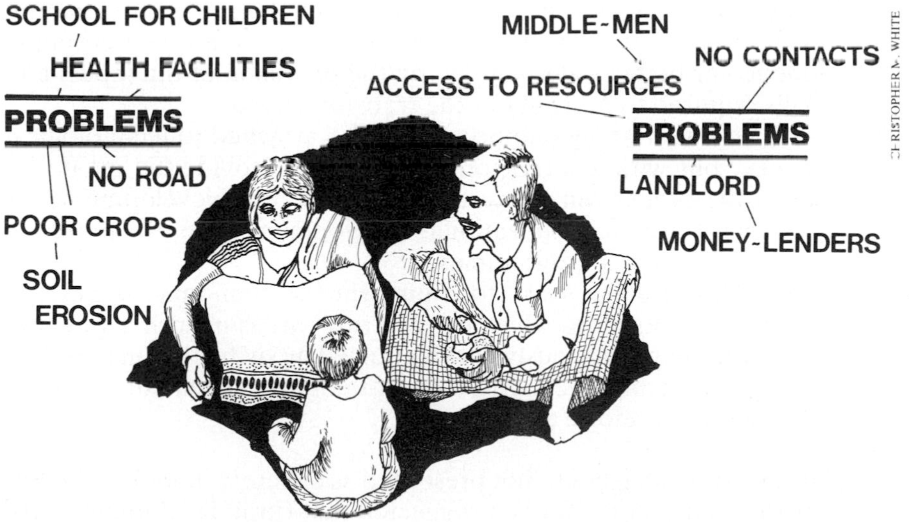

Farmers and their families face a whole range of problems

In thinking of rural development, therefore, a whole range of problems which the farmer confronts daily must be considered. Some of these problems will be physical or tangible, and relatively easy to identify. They can quickly be spotted by observation or by means of a survey and once the extent of the problem is understood a relevant course of action can be proposed. For example, fertilizer can be recommended to improve the production level of a certain crop.

However, not all of the problems that farmers face are physical nor can they always easily be seen. Many of these problems derive from the farmer's place in the social and political structure in the rural area. Farmers and their families are involved in a complex web of relationships with other farmers in the area and often these relationships bring about problems. Dependence upon a money-lender, for example, is a problem facing many farmers in developing countries. Farmers may also have little access to the resources necessary for development, nor any way of getting such resources. Finally, they may have had very little contact with rural development programmes or other government services, and may not know how to take advantage of such activities.

It should be emphasized that the problems a farmer faces are complex and not all of them are physical or tangible. With this in mind, the kinds of strategies which rural development programmes can adopt can be considered. The first point to make is that there is no one strategy which is relevant to the problems of all rural areas. Different areas have different kinds of problems and the strategy must be adapted accordingly. There are three broad rural development strategies to be considered.

Technological. Here, the emphasis is upon technological transformation of different aspects of the rural society, e.g., improved cropping practice or better water supply, by the provision of the inputs and skills required to bring about the transformation.

Reformist. In this strategy, importance is also attached to technological change, but with a corresponding effort to provide the means by which the farmer can play a bigger part in rural development, for example, through organizational development, or participation in rural development programmes.

Structural. This strategy seeks to transform the economic, social and political relationships which exist in rural areas in such a way that those who were previously disadvantaged by such relationships find their position improved. Often this strategy is carried out by means of an agrarian reform programme.

The above strategies are not presented as concrete models to be followed without question. Nor is it suggested that rural development programmes must adopt any one strategy. They are presented to show the range and mixture of strategies which a rural development programme can follow. A farmer's problems will probably demand different action at different levels if they are to be tackled in a comprehensive manner.

Principles of rural development programmes

Rural development strategies usually take the form of programmes which implement projects in a specific rural area. Such programmes form the basis of most government and non-government efforts to assist rural areas, and they include both agricultural and non-agricultural projects, e.g., maternal and child health programmes. Specialized staff supply the expertise required, and ministerial or other institutional budgets provide the necessary financial resources. External aid is also usually channelled into such programmes in the rural areas.

While this guide does not intend to examine the areas of programme planning or implementation, it does suggest a number of very broad principles which should be followed by rural development programmes. The content of these programmes is a matter for the specialists in the particular field, i.e., agriculture, health or water supply. It is important, however, for all such programmes to establish beforehand a set of principles to guide their activities. The following principles are suggested to implement rural development programmes.

Access. Try to ensure that the programme and its benefits can reach those in need, and beware of the consequences if some farmers have access to the programme while others do not.

Independence. Devise a programme which helps and supports the farmer but which does not make him or his livelihood dependent upon the programme.

Sustainability. Ensure that the programme's plans and solutions are relevant to the local economic, social and administrative situation. Short-term solutions may yield quick results, but long-term programmes that are suitable to the local environment have greater success.

Going forward. Technological aspects of rural development programmes should help the farmer to take the next step in his development and not demand that he take a huge technological leap. It is better to secure a modest advance which can be sustained than to suggest a substantial advance which is beyond the ability of most.

Participation. Always try to consult the local people, seek out their ideas and involve them as much as possible in the programme.

Effectiveness. A programme should be based on the effective use of local resources and not necessarily on their most efficient use. While efficiency is important, its requirements are often unrealistic. For example, the maximum use of fertilizer is beyond the means of most farmers. But an effective use of resources, which is within the capabilities of most farmers, will have a better chance of a wider impact.

The importance of extension

Within the framework presented in this chapter, the concept and practice of the central issue of this guide must now be examined: extension work in rural communities. Extension is essentially the means by which new knowledge and ideas are introduced into rural areas in order to bring about change and improve the lives of farmers and their families. Extension, therefore, is of critical importance. Without it farmers would lack access to the support and services required to improve their agriculture and other productive activities. The critical importance of extension can be understood better if its three main elements are considered:

KNOWLEDGE ↔ COMMUNICATION ↔ FARM FAMILY

Extension is not concerned directly with generating knowledge; that is done in specialized institutions such as agricultural research centres, agricultural colleges or engineering departments. Extension takes this knowledge and makes it available to the farm family. Rural extension, therefore, is the process whereby knowledge is communicated, in a variety of ways, to the farm family. This process is usually guided and supported by an extension agent who works at the programme and project level, and who is in direct contact with farmers and their families.

To do this extension work, agents have to be trained in the different aspects of the extension process. One aspect of this training is giving the agent the technical or scientific knowledge required for the job. This is usually done during the agent's professional training; however, it is only one element in the process. The other two elements of the process are equally important. It is not enough for an extension agent to have technical knowledge; he must also know how to communicate this knowledge and how to use it to the benefit of the farm family. Training in extension, therefore, is an equally important aspect of the training of any agent who wishes to work with farmers.

2. Understanding extension

The concept and practice of extension are the central themes of this guide. However, before beginning to look at the many different aspects of extension practice in later chapters, the meaning of the term *extension* needs to be examined. Rural extension is now a common activity in most countries of the world, and it is a basic element in programmes and projects formulated to bring about change in rural areas. Extension services are similarly a common feature of the administrative structure of rural areas and these services have the responsibility, in partnership with the farmers, of directing programmes and projects for change.

The concept of extension

Extension is a term which is open to a wide variety of interpretations. Each extension agent probably has his own understanding of what extension is. This understanding will be based on past experience and the particular type of extension service in which the agent is working. In other words, there is no single definition of extension which is universally accepted or which is applicable to all situations. Furthermore, extension is a dynamic concept in the sense that the interpretation of it is always changing. Extension, therefore, is not a term which can be precisely defined, but one which describes a continual and changing process in rural areas.

The term extension may be examined by looking at a number of statements that have been written about it.

- Extension is an informal educational process directed toward the rural population. This process offers advice and information to help them solve their problems. Extension also aims to increase the efficiency of the family farm, increase production and generally increase the standard of living of the farm family.
- The objective of extension is to change farmers' outlook toward their difficulties. Extension is concerned not just with physical and economic achievements but also with the development of the rural people them-

selves. Extension agents, therefore, discuss matters with the rural people, help them to gain a clearer insight into their problems and also to decide how to overcome these problems.

- Extension is a process of working with rural people in order to improve their livelihoods. This involves helping farmers to improve the productivity of their agriculture and also developing their abilities to direct their own future development.

The above statements are presented to illustrate the range of interpretations that can be found about extension. They do, however, contain a number of common points. They all stress that extension is a process which occurs over a period of time, and not a single, one-time activity. They also all underline extension as an educational process which works with rural people, supports them and prepares them to confront their problems more successfully.

If statements such as those above are examined more carefully, and if the current ideas and practice of extension are considered, four main elements can be identified within the process of extension: knowledge

F. BOTTS

Extension is an educational process. Here, an extension worker discusses with a Thai farmer ways of improving his land by terracing and planting legumes

and skills, technical advice and information, farmers' organization, and motivation and self-confidence.

Knowledge and skills

Although farmers already have a lot of knowledge about their environment and their farming system, extension can bring them other knowledge and information which they do not have. For example, knowledge about the cause of the damage to a particular crop, the general principles of pest control, or the ways in which manure and compost are broken down to provide plant nutrients are all areas of knowledge that the agent can usefully bring to farmers.

The application of such knowledge often means that the farmer has to acquire new skills of various kinds: for example, technical skills to operate unfamiliar equipment, organizational skills to manage a group project, the skill to assess the economic aspects of technical advice given, or farm management skills for keeping records and allocating the use of farm resources and equipment.

The transfer of knowledge and skills to farmers and their families is an important extension activity and the extension agent must prepare himself thoroughly. He must find out which skills or areas of knowledge are lacking among the farmers in his area, and then arrange suitable learning experiences through which the farmers can acquire them.

Technical advice and information

Extension also provides advice and information to assist farmers in making decisions and generally enable them to take action. This can be information about prices and markets, for example, or about the availability of credit and inputs. The technical advice will probably apply more directly to the production activities of the family farm and to the action needed to improve or sustain this production. Much of this technical advice will be based upon the findings of agricultural research. In many instances, however, farmers are also sources of valuable advice and information for other farmers, and agents should always try to establish a farmer-to-farmer link.

Farmers' organization

As well as knowledge, information and technical advice, farmers also need some form of organization, both to represent their interests and to give them a means for taking collective action. Extension, therefore,

CHRISTOPHER M. WHITE

Extension brings low-cost improvements in farming practice, such as line planting of crops

should be concerned with helping to set up, structure and develop organizations of local farmers. This should be a joint venture and any such organization should only be set up in consultation with the farmers. In the future, these organizations will make it easier for extension services to work with local farmers, and will also serve as a channel for disseminating information and knowledge.

Motivation and self-confidence

One of the main constraints to development that many farmers face is isolation, and a feeling that there is little they can do to change their lives. Some farmers will have spent all their lives struggling in difficult circumstances to provide for their families with little support or encouragement. It is important for extension to work closely with farmers, helping them to take the initiative and generally encouraging them to become involved in extension activities. Equally important is to convince farmers that they can do things for themselves, that they can make decisions and that they have the ability to break out of their poverty.

The above are the four fundamental elements of the extension process. It is not suggested that all extension activities must contain each of these elements, nor that some are more important than others. Clearly, the extension approach will be determined by the particular circumstances. However, an overall extension service should be based on these elements and should seek to promote them within the rural areas. Sometimes the local farmers' problems will demand prompt information and advice; on other occasions, more patient work of organization and motivation may be required. An extension service must be able to respond to these different demands.

Principles of extension

Extension activities are widespread throughout the developing world and most governments have set up formally structured extension services to implement extension programmes and projects. The practice of extension is supported by budget, offices, personnel and other resources. Before examining extension in detail in later chapters, however, it will be useful to consider the principles which should guide it.

Extension works with people, not for them

Extension works with rural people. Only the people themselves can make decisions about the way they will farm or live and an extension agent does not try to take these decisions for them. Rural people can and do make wise decisions about their problems if they are given full information including possible alternative solutions. By making decisions, people gain self-confidence. Extension, therefore, presents facts, helps people to solve problems and encourages farmers to make decisions. People have more confidence in programmes and decisions which they have made themselves than in those which are imposed upon them.

Extension is accountable to its clients

Extension services and agents have two sets of masters. On the one hand, they are accountable to their senior officers and to the government departments that determine rural development policies. Agents are expected to follow official policies and guidelines in their work.

On the other hand, extension is the servant of the rural people and it has the responsibility to fulfil the needs of the people in its area. This means that the rural poor should have a say in deciding how effective extension actually is. One measure of effectiveness is to see how well policies and plans have been carried out. An equally important measure is the extent to which incomes and living standards of the rural people have increased as a result of extension work.

Extension programmes, therefore, are based on people's needs, as well as on technical and national economic needs. The extension agent's task is to bring these needs together. For example, an important part of government policy may be to increase the amount of food grown and sold in the country. By choosing to encourage the mass of small farmers to increase their output by improving their farming methods, national needs and farmers' needs can be satisfied together.

Extension is a two-way link

Extension is not a one-way process in which the extension agent transfers knowledge and ideas to farmers and their families. Such advice, which is often based upon the findings of agricultural and other research stations, is certainly important but the flow of information from farmers to extension and research workers is equally important. Extension should be ready to receive farmers' ideas, suggestions or advice, as well as to give them. This two-way flow of ideas can occur at different stages.

When the problem is being defined. Being in regular contact with the farmers, the extension agent can help research workers to understand the farming problems of the area and the limitations under which farmers

CHRISTOPHER M. WHITE

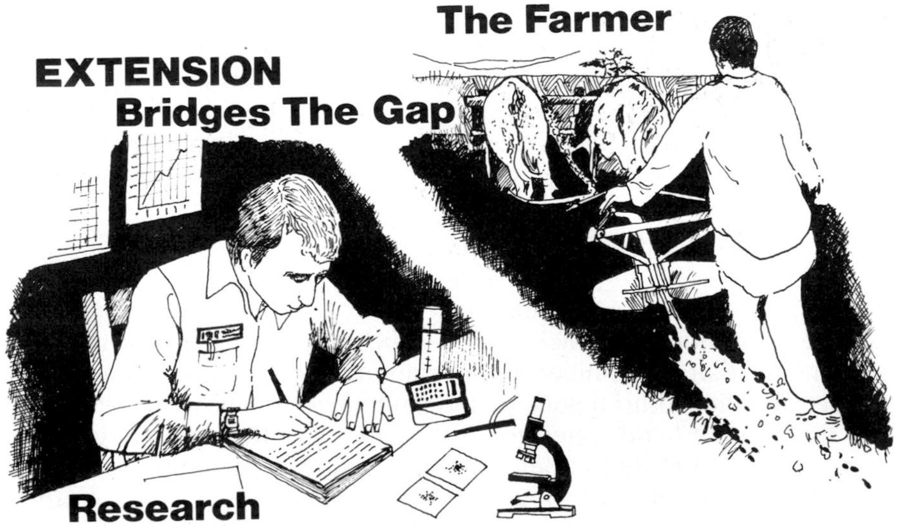

Extension links farmers with research

have to work. It is even better if the agent can bring researchers into direct contact with farmers in order to ensure that research recommendations are relevant to farmers' needs.

When recommendations are being tested in the field. A new farm practice or crop variety might produce good results at a research station but not do so well on a farmer's field. Trials on farmers' fields are an opportunity to test research recommendations and provide feedback for research staff.

When farmers put recommendations into practice. Sometimes farmers discover problems with a recommendation which the research station failed to note. With the feedback the recommendations can be adjusted accordingly.

The two-way link between research, extension and the farmer is fundamental to sound extension practice and should be a basic principle of extension activity.

Extension cooperates with other rural development organizations

Within rural areas, extension services and agents should work closely with the other organizations that provide essential services to farmers and their families. Extension is only one aspect of the many economic, social and political activities that seek to produce change for the better in rural society. Extension, therefore, must be prepared to collaborate with all other such organizations, both government and non-government, and to take them into account when preparing to implement extension policies. The kinds of organizations with which extension services should cooperate include:

Political institutions and local political leaders whose active local support will help the extension agent, who may thereby be brought into closer touch with local farmers.

Support organizations such as those which supply agricultural or other inputs, credit facilities or marketing services. Such inputs must be available in sufficient quantity, in the right place and at the right time if they are to be of any use.

Health services, so that the extension agent is kept aware of local health problems, particularly nuritional levels. Agricultural development and nutrition are closely related and the agent must keep closely in touch with health programmes and projects and adapt his programme to conform to local health requirements.

Local schools, so that the agent can have early access to the farmers of the future, and begin to equip them with the knowledge and skills required for farming.

Community development, whose objectives will be very similar to the educational work of extension. Extension agents often work very closely with community development workers to break down local social and cultural barriers to change, and to encourage community action programmes.

It is essential that the extension agent in the field know what his colleagues in other services and government departments are doing, and that they understand what he is doing. Close cooperation not only avoids duplication but provides opportunities for integrated farm programmes.

Extension works with different target groups

Extension recognizes that not all farmers in any one area will have the same problems. Some will have more land than others and will be keen to try out new ideas. Others, with fewer resources, will probably be more cautious. Extension cannot offer a single "package" of advice, suitable to all farmers. Different groups need to be identified and the agent will have to develop programmes appropriate to each group.

In the past, much extension effort was concentrated on the progressive farmer who was expected to spread new ideas to others. It has been seen, however, that this does not always work, because progressive farmers often have different problems. They have more land, more education and are usually more involved in the marketing of their produce.

Extension must, therefore, be aware of the existence of different farming groups and plan its programmes accordingly. The smallest and poorest farmers will need particular attention, as they may lack the basic resources needed to become involved in extension activities. The point to stress, therefore, is the existence of farmer groups with different resources and skills in any one community, and the need for extension to respond to these groups accordingly.

Extension and education

It has been seen that the extension agent's task is an educational one. Farmers and their families need to learn new skills, knowledge and practices in order to improve their farming and other productive activities. As they do so, they develop new attitudes toward farming and the new practices, and to extension itself; this in turn influences their future behaviour. Extension agents, however, must also be prepared to learn from farmers about the way they farm, and keep themselves up to date with relevant developments in agricultural knowledge. In this educational work of extension, the agent should be aware of a number of principles of learning.

The educator must also be a learner

Education is not a process of filling empty minds with knowledge. Farmers already have a lot of knowledge about their environment and about their farming system; they would not be able to survive if they did not. Extension must build on the knowledge that already exists.

An extension agent, therefore, needs to learn as well as to teach. He must learn what farmers already know about agriculture: for example, how they describe and explain things that happen on their farms and what they know already about improved farming methods.

Learning requires motivation

No one can compel another person to learn. There has to be a desire to learn. Adults find it more difficult than children to grasp new ideas and information. Also, unlike a schoolteacher, the extension agent does not have a captive audience. Farmers can choose not to learn and they can choose not to listen to extension agents.

People do not learn unless they feel that the learning will result in their being able to satisfy a need or want. Food and drink are needs that are essential for life, together with the starting and raising of a family and the search for safe living conditions, which provide protection and shelter from danger and discomfort for the family.

In addition to needs, people have wants or desires. These are less intense than needs, but still important. People desire approval and praise from their family and friends; they want prestige in their society and to be well thought of by their neighbours. These desires become more apparent once basic needs have been satisfied. Farmers and their families who are wellfed and have good homes still strive for improvement. They want to produce more and the extension agent, by helping them to improve farming methods, can use this legitimate ambition to help them to improve the productivity of their farms. A farmer who is motivated to learn is likely to do so more rapidly and completely than a farmer who lacks motivation. This is a very important principle for the extension agent to remember.

GOVERNMENT OF INDIA

Practice is important when learning new skills. Here, members of a Young Farmers' Club in Sisarma, India, gain practical experience in maintenance of a mechanical water pump

Dialogue and practice are important for learning

An extension agent tells a group of farmers how to thin their crops in order to improve yields. He then goes away, thinking that the farmers have learned the new skill. A few weeks later, he returns to find that none of the farmers have thinned their crops and that they have only a very vague idea of what he told them.

The extension agent should not be surprised. Farmers do not learn very much from a straightforward talk and most of what they do hear they soon forget. But if they are given the chance to ask questions, to put the new information into their own words and to discuss it with the extension agent, much more will be learned and remembered. Furthermore, when a new practical skill is being taught, the farmers must have a chance to practise it. The extension agent can then correct any initial mistakes, and the farmer will gain the confidence to use the new skill.

Learning and adoption occur in stages

Different types of learning are involved in extension. Before a group of farmers can decide to try out a new practice, they must first learn of its existence. They may then have to learn some new skills. Five stages can be identified in the process of accepting new ideas.

Awareness. A farmer learns of the existence of the idea but knows little about it.

Interest. The farmer develops interest in the idea and seeks more information about it, from either a friend or the extension agent.

Evaluation. How the idea affects the farmer must now be considered. How will it be of benefit? What are the difficulties or disadvantages of this new idea? The farmer may seek further information or go to a demonstration or meeting, and then decide whether or not to try out the new idea.

Trial. Very often, farmers decide to try the idea on a small scale. For example, they may decide to put manure or fertilizer on a small part of one field and compare the result with the rest of the field. To do this they seek advice on how and when to apply fertilizer or manure.

Adoption. If the farmers are convinced by the trial, they accept the idea fully and it becomes part of their customary way of farming.

Similar stages are involved with individual farmers, farmers' groups, or whole communities. In groups and communities, the process is more complex and may take much longer. The extension agent uses a range of extension methods to bring the right kind of information and support to each stage of the process. He must arrange learning experiences that will lead people from one stage to the next. In a community forestry programme, he begins by encouraging people to recognize that there is a problem of declining numbers of trees and that this could be overcome by the community planting and looking after a wood-lot. Interest can then be increased by a visit to another village that has already planted a wood-lot. During the evaluation stage, a lot of discussion will go on in the village. The extension worker can provide detailed information about the costs and returns, and answer questions and doubts. When a decision is taken to do something, he can then arrange skill training sessions.

Farmers differ in their speed of learning and adoption

The process by which a new idea spreads among people in an area is known as diffusion. Not all farmers will accept a new idea at the same time. In any rural community, the readiness to accept new ideas and put them into practice varies from farmer to farmer depending on each farmer's previous experience with new ideas, the personality of the farmer and the amount of land and other resources available. Thus we can identify different categories of farmers in terms of their abilities to adopt new ideas.

Innovators. Innovators are farmers who are eager to accept new ideas. Usually there are only a few people in this class in a farming community. They are often farmers who, having spent some years outside the village, feel that they can make their own decisions without worrying about the opinions of others. In villages, innovators are often looked on with suspicion and jealousy. Yet they are important to the success of an extension programme since they can be persuaded to try new methods and thereby create awareness of them in the community. However, the extension agent should exercise tact and caution, and avoid overpraising innovators in public or spending too much time with them. This could result in rejection of the idea by the rest of the community because of jealousy and suspicion of the innovator's motives in adopting new methods.

Early adopters. Farmers who are more cautious and want to see the idea tried and proved under local conditions are known as early adopters. They express early interest but must first be convinced of the direct benefit of the idea by result demonstration. Usually this group of farmers includes local leaders and others who are respected in the community.

The majority. If the rest of the farmers adopt a new idea, they will do so more slowly and perhaps less completely. Many farmers will lack the resources to adopt the new idea at all, while others may only do so slowly and with caution. The majority who can and do adopt the idea are likely to be more influenced by the opinions of local leaders and neighbours than by the extension agent or the demonstrations he arranges.

Types of extension

There is no one universal type of extension but a variety of activities and approaches which can be called extension. It has already been stated that since agriculture is the basis of a rural economy, agricultural extension is the most common type of extension to be found in rural areas. But the areas of knowledge and new ideas that farmers and their families require are not restricted to agriculture. There are other aspects of family life in which new knowledge and practices can lead to improvement. Extension is any activity that works with farmers and their families in order to improve the economic and social conditions of their lives and to develop their ability to take responsibility for their own future development. This extension, however, can take different forms and it would be useful to review the two principal ones.

Agricultural extension

There are probably more extension agents involved in agricultural activities than in any other aspect of rural life. Given the importance of agriculture and the need to produce food both for the farm family and for the nation as a whole, this emphasis upon agricultural extension is understandable. Some agricultural extension services are based upon a single crop, while others adopt more of a "whole farm" approach. The choice is very much dependent upon the local agricultural system and the national crop requirements. In regions where cash crops such as cotton, cocoa or sugar grow, the single-crop extension approach is more common.

An agricultural extension service offers technical advice on agriculture to farmers, and also supplies them with the necessary inputs and services to support their agricultural production. It provides information to farmers and passes to the farmers new ideas developed by agricultural research stations. Agricultural extension programmes cover a broad area including improved crop varieties, better livestock control, improved water management, and the control of weeds, pests or plant diseases. Where appropriate, agricultural extension may also help to build up local farmers' groups and organizations so that they can benefit from extension programmes. Agricultural extension, therefore, provides the indispensable elements that farmers need to improve their agricultural productivity.

Non-agricultural extension

In the absence of a collective term to cover the other types of extension, it is convenient to refer to them all as non-agricultural extension. This term includes all activities and efforts not directly related to agriculture or livestock production, but which are important to the farm families. Home economics, family health and nutrition, population education and community development are all non-agricultural extension activities.

Rural extension covers many aspects of rural life

When talking of extension and extension agents, therefore, all activities of the above type are included. These activities also involve the basic elements and principles of extension outlined earlier in this chapter, such as knowledge, learning and practice. Home economists and community development workers, therefore, are extension agents who deal with farm families in the same way as agricultural extension agents. The only difference is their areas of concern.

In fact, it is becoming increasingly common to talk of rural extension as a collective term which brings together all agricultural and non-agricultural extension activities. The feature common to both types of extension is that they work with families in rural areas and deal with problems in a rural environment. Their different programmes and approaches have a common aim, which is the improvement of the lives of the rural people, and they are both guided by common principles and ideals.

This guide, therefore, is a guide to rural extension and is relevant to both agricultural and non-agricultural extension agents. Given the predominance of agriculture in the rural economy, however, there will be some emphasis on agricultural extension within the guide. The understanding of extension, the methods used by extension agents, the planning processes involved and the qualities and skills required by agents are factors relevant to all forms of rural extension. The content and subject-matter may be different, but the same general principles apply in both types of extension.

3. Social and cultural factors in extension

Farmers and their families are members of the society in which they live. In any society there are strong pressures on its members to behave in certain ways. For the farmers, some of these pressures will come from within. In all societies there are accepted ways of doing things and these ways are directly related to the culture of the society. Farmers' attitudes and desires are influenced by their society's culture. If it is customary in a certain community for farmers to scatter seed and plough it into the soil, people will grow up to believe that that is the only correct way of planting. Even if the benefits of other methods are explained to them, their strongly held attitudes may make it difficult for to them change.

Yet not all of these pressures will come from the farmers' own attitudes and beliefs; some will come from other people. Any society expects its members to behave in certain ways. No one is seen by others as an isolated individual. Each person is seen as occupying a position in society, and each position carries expectations with it. In some communities, an unmarried man is expected to work on his father's farm; only when he marries will people expect him to start farming his own plot. A successful farmer may be expected to give food, money and shelter to relatives who have not been so successful, or to pay for his relatives' children to go to school. If a person resists these expectations, those around him will show their disapproval. Because most people like to feel acceptance and approval from those around them, they tend to behave in accordance with such expectations.

An extension agent will be more effective if he understands the social and cultural background of the farmers with whom he works. He will then be better able to offer advice that fits in with the culture of the society, and he can use the structure and culture of the society to the benefit of his work. It is useful, therefore, to examine the main features of societies and cultures that are relevant to extension work.

Social structure

The structure of a society is the way it is organized into families, tribes, communities and other groupings or divisions. A person's attitudes, and people's expectations of that person, are influenced by the groups to which he or she belongs; so too is the individual's access to opportunities, jobs and land.

Social divisions

Divisions within a society can be based on several different factors, including age, sex, religion, residence, kinship and common economic interest.

Age

People of the same age usually have similar interests and attitudes. Young people tend to have different values, attitudes and aims in life from those of older people. In many societies, elderly people are treated with great respect, and their advice is listened to carefully. An extension agent needs to learn the particular aims, expectations and restrictions of different age groups in the society in which he works.

Sex

Traditionally, in rural areas, specific tasks are done either by men or women. Usually women are responsible for household jobs, such as cooking, collecting water and firewood or looking after children. However, in many countries, women also do a lot of farm work. In a number of African countries, over 60 percent of all agricultural work is usually done by women. Often, women have their own fields in which they grow food crops, while the men are responsible for commercial cash crops such as tobacco or oil-palm.

Elsewhere, men and women work the same fields, but carry out different tasks. In Botswana, for example, ploughing and all work connected with cattle are traditionally a man's job, while weeding, bird-scaring and threshing are done by the women. Agricultural extension often concentrates on men, with male extension agents visiting male farmers. But any change in the way people farm will also affect the women, and thus may well fail unless extension agents involve women in their programmes.

Religion

Members of religious groups have common beliefs and attitudes, and these may influence their willingness to work closely with people of other religions. Religious differences can create tensions in a rural community;

CHRISTOPHER M. WHITE

Social division of labour: men and women have different responsibilities on the farm

the extension agent should be aware of these. Some religions impose patterns of behaviour which may affect extension. Certain times of day, particular days of the week or seasons of the year may be devoted to religious ceremonies, which means that farmers are not available for farm work or for extension activities.

Residence

People who live close to one another usually have some interests in common. Residents of a village will want facilities such as a school, clean water and health services. They will want access to roads and a fair share in government development programmes. These common interests can unite the village, particularly if such interests are threatened. Where possible, extension agents should try to include in their programmes activities which will unite the whole community in a common task. But they should be aware that there may also be divisions within a village. For example, residents of one part of the village may want a new water tap to be put near their homes, while others will argue that it should be near them.

Where there is tension between different parts of a community, extension agents should as far as possible avoid making it worse and, wherever possible, they should seek ways to reduce this tension. If an agent is seen to be working on behalf of one particular group in the village, other groups may make it very difficult for him to be effective.

Kinship

The strongest groupings are often those based on relationships of birth and marriage within and between families. The smallest of these groupings is the family, which consists of a man and woman and children. In some societies, such families are independent and make their own decisions about where to live, where to farm and what crops to grow. These families will, however, usually have certain duties toward close relatives that they will be expected to fulfil, and these could restrict their freedom of action.

In other societies, larger kinship groups may live together, own land in common or even take joint decisions about farming. When this happens the individual farmer may have little freedom of decision. An extension agent would need to find out who are the leaders and decision-makers of such groups, and work closely with them.

Common economic interest

Economic differences are an important part of social structure. The type of job people do, the amount of money they earn and the quality of land they own or can rent are factors which can divide society into distinct groupings, each with its own concerns, interests, values and attitudes. In a rural area, there may be cattle owners and crop farmers; subsistence farmers who cannot afford to buy costly inputs; commercial farmers who are interested in learning about the latest farm equipment; and landowners and tenants. Each group will have its own requirements and expectations of the extension agent, and the agent will need to adjust his approach to each group's interests.

The most important economic factors creating divisions within rural societies are the amounts of land and money that each farmer has. This is particularly clear in some Latin American countries where a small proportion of families own very large estates while most families work as farm labourers or farm their own very small plots on a subsistence basis. Most Asian countries also have large numbers of landless labourers as well as small and large farmers. Each of these categories of farmers has very different needs in terms of extension support.

Economic differences affect the type of advice and support that an extension agent should offer each category of farmer. Such differences determine the standard of living that people can achieve and they also affect a farmer's relative economic and political influence. Large farmers are more likely to be given credit than small farmers, and merchants and

Large farmers and small farmers need different kinds of help from extension

traders will give them better terms because they buy and sell in larger quantities. Planners and political leaders often listen to them more readily. Extension agents may also find it more attractive to work with the larger farmers. But if smaller farmers are to be helped, extension agents should be aware of these divisions and look for ways of supporting those farmers who are keen to improve their farms but have not much political or economic influence.

Groups

The broad social divisions that affect the attitudes, needs and interests of the members of a society have been discussed. There are also, in all societies, small groups of people who come together for a common purpose or activity. Some of these groups may stay in existence for a long time. A savings club, for instance, may continue to meet week after week for many years. Although the members may change, the club will remain. Other groups may be temporary, such as when several neighbours agree to help with the farm work on each other's land.

These groups can be very useful for extension agents and can often form the basis of extension groups. In the Republic of Korea, for example, traditional women's savings groups have developed into Mothers' Clubs, which are extremely influential in village development activities. These clubs raise large sums of money for community projects, contribute labour for self-help projects, and are a channel for information on farming and popular education for rural women.

People, however, vary in their readiness to join groups. In some communities, for example, kinship groups may own land jointly but leave each small family to farm its own plots. In this situation, any change in farming practice would depend on separate decisions by many individual farmers. Elsewhere, the members of the kinship groups may farm the land together. In that case, an extension agent should work with the group as a whole, probably through its traditional leaders, to improve farming practice.

An extension agent should use such groups whenever possible, and to assist his work he should try to find out (*a*) what groups there are in the region, (*b*) what joint activities are undertaken by members of each group, (*c*) the interests of each group, and (*d*) whether any of the groups could form the basis of an extension group.

Formal and informal leaders

In all societies there are men and women who make decisions on behalf of others, or who are respected by others, and therefore have some influence on their attitudes and behaviour. Such leaders can be very important for the success of extension work.

People who hold recognized positions of authority are known as formal leaders. They are usually easy to identify once the pattern of leadership in the society is understood. Some inherit their position; others are elected, and others are appointed by someone in higher authority. Leadership may be shared by several people or be held by a single person. In most social communities there are religious leaders, such as priests, as well as secular leaders, such as elected councillors and village heads.

In any rural community there will be a number of formal leaders: for example, religious leaders; the chairman of a cooperative; a traditional headman supported by an advisory group of elders; heads of kinship groups and families; a village development committee; local leaders of political parties; or elected councillors. The exact pattern will vary from one society to another, but the extension agent should learn what the role of each leader is, and how much influence each has within the community. A village headman, for example, may have the power to allocate land to farmers who want to expand their holdings. In this situation, the extension agent will need the headman's support if he is to encourage farmers to invest in new enterprises which require additional land.

Extension agents should try to work through formal leaders. They must learn which person is the best to approach on a particular issue. This may vary from place to place, even within an extension agent's area. A traditional chief in one village may be more influential than an elected councillor, while in a neighbouring village the opposite may be the case.

In many rural societies, the extension agent will have little success unless he first gains the support of the traditional leaders. Only then will he be able to win the trust and confidence of the members of the community.

Informal leaders are not so easy to identify, because they do not hold any particular position of authority. They are individuals who are respected by other people, not because they hold an official position but because they have an attractive or forceful personality or because they seem to know the best action to take in any situation. Whatever the reason may be, other people are influenced by them. If informal leaders in a community support a new idea, such as the planting of a village wood-lot or the setting up of a cooperative, then others will be more ready to support it. Extension agents can find out who these influential people are by observing who speaks out at village meetings or by asking farmers who they normally go to for advice.

An extension agent can be more effective if he works through the existing structure of a rural society and through its formal and informal leaders. However, such an approach also has its limitations. Influential leaders often come from the more privileged sections of the community. They may simply keep the benefits of extension, and of agricultural credit and inputs, to themselves and their friends. By working through such leaders, extension may widen the gap in living standards between the different sections of society. The agent, therefore, should seek to work through existing formal and informal leaders, but should ensure that this approach does not leave some farmers at a disadvantage.

Social expectations

It was stated earlier that a person's position will determine the way others expect him or her to behave. These expectations are known as norms. It is the norm in some societies, for example, for a married woman to eat her meal only after her husband has finished eating. These norms are deeply ingrained in people's attitudes and beliefs. They not only determine how other people think an individual should behave; they determine what behaviour the individual feels is correct. Extension agents should be sensitive to these expectations and should not underestimate their influence on people's behaviour, however irrational they may seem at first.

Culture

The culture of a society is the accepted way of doing things in that particular society. It is the way in which people live, their customs, traditions, methods of cultivation and so on. The culture of a society is learned by each individual member of that society. Children are not born with this knowledge; they learn by seeing how older children and adults behave. As they grow up, older members of their family or kinship group teach them about the customs and traditions of the group and the society. Later still, they may be initiated more fully into the society at ceremonies where they are taught traditional habits and customs, and their expected role. Experience also gives the individual a better understanding of the behaviour pattern of the community and may teach the individual how to change some of the traditional forms of behaviour for newer, more modern forms.

Culture is not an accidental collection of customs and habits but has been evolved by the people to help them in their conduct of life. Each aspect of the culture of a society has a definite purpose and function and is, therefore, related to all the other aspects of its culture. This is important to remember when planning extension programmes. Changes in one aspect of culture may have an effect on other aspects of that culture. If changes in one aspect of culture are introduced, and these are likely to have an unacceptable effect on other aspects, then a programme may have little chance of success. This is one reason why local leaders and farm people should help in planning an extension programme. They will know whether or not the changes proposed will be acceptable to the society.

The more an extension agent learns about and comes to respect the culture of the people with whom he works, the more he will be accepted by them. He will also be more sensitive to the type of advice and support that will be useful.

There are five particular aspects of local culture that the extension agent should be aware of: the farming system, land tenure, inheritance, ceremonies and festivals, and traditional means of communication.

Farming systems

Before he can offer any advice to farmers, the extension agent must understand their present farming system. What crops are grown and in what sequence or combination? How important is each crop in the local diet? How is land prepared for planting? When are the main farm operations carried out? Why do people farm in the way they do? Farming systems are complex, and change in one aspect may create problems in others. In parts of Nepal, for example, millet is sometimes planted between maize plants. Thus, any change in maize spacing or subsequent weeding practice will affect millet production. Similarly, in regions of Nigeria, up to 12 crops may be grown together on a single plot.

Once he is familiar with local farming systems, the extension agent can explore the possibilities for improvement. New farming practices will be more acceptable to farmers if they can be introduced into existing systems without drastic changes. Perhaps the timing of certain operations can be adjusted, or weeding carried out more regularly. Different seed varieties could be tried, or water use improved to provide more irrigated land. It is important to begin with what is already there and build upon it.

Farming practice is not isolated from the rest of the society's culture and it cannot be treated as a purely technical subject. It influences, and is influenced by, other aspects such as food preferences, land tenure and family relationships. In one African country, for example, extension agents encouraged farmers to plant their crops a few weeks earlier than they usually did. Research findings showed that output would increase and that even if the early sowing failed because of lack of rain, farmers would have the chance of re-planting. However, this advice challenged the authority of traditional leaders. Nobody was supposed to begin ploughing and planting until the village headman had declared that the time was right. The advice also conflicted with the relationship between cattle owners and arable farmers: cattle were allowed to graze freely on the stubble and grass in the fields until the planting season began. This simple recommendation, therefore, had implications for other aspects of culture, which made it difficult for individual farmers to change their farming practice.

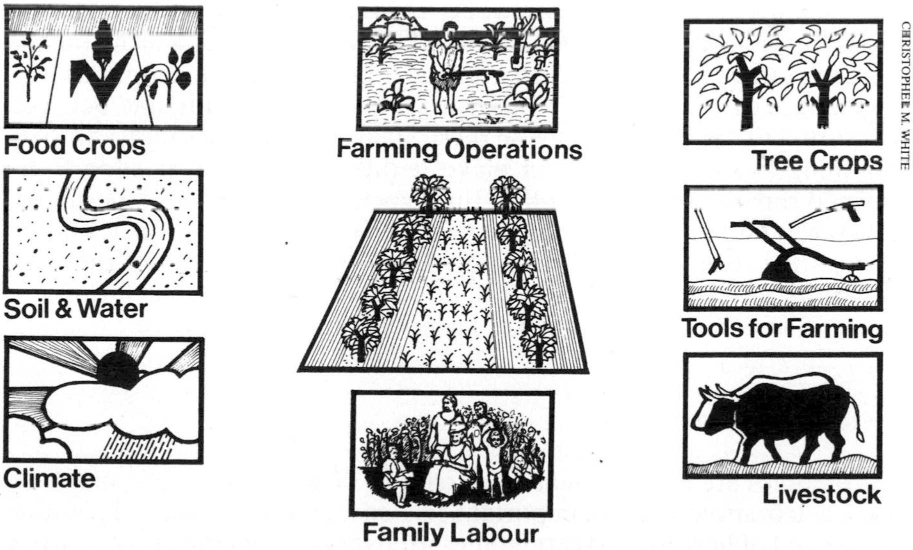

Farming systems: change in one part of a complex system will affect other parts

Land tenure

Land tenure consists of the ways in which people obtain the right to possess and use land. Land-tenure systems vary from one society to another. In some communities land is owned by a tribe or kinship group, and each family has the right to use as much land as it needs to feed itself. It cannot sell or rent that land to anyone else, and there may be restrictions on the uses to which the land can be put. In other societies individuals can buy land and do what they like with it.

The land-tenure system will affect people's ability and incentive to take extension advice. In some countries, for example, land is farmed on a share-cropping basis. The farmer gives a fixed proportion of everything that is produced on the land to the landowner. The farmer will, therefore, be unwilling to adopt new practices if most of the benefits will go to the landowner. Elsewhere, a young farmer may want to plant a tree crop, but is not allowed to do so by the leaders of the kinship group that owns the land. Or perhaps a tenant would like to improve his farm by fencing it or installing an irrigation pump but may decide not to, fearing that his landlord may take back the land without paying him any compensation for the improvements.

Inheritance

The way in which land and other possessions pass from one generation to the next also affects extension work. In some cultures, a man's possessions are inherited not by his children but by his mother's brothers and their children. This may reduce a farmer's incentive to develop the farm. In many areas, it is normal practice for a man to divide his land between his sons and daughters before he dies. Such a farmer will not want to do anything to the land that will make it difficult for each portion to be farmed separately later. In other rural societies, land is not inherited at all. When farmers die, the land they farmed is taken back by their kinship groups for reallocation. Extension agents should understand the local inheritance rules, because they will affect the ability of young farmers to acquire land, and the incentive of farmers to take their advice.

Ceremonies and festivals

Ceremonies are a central feature of culture. They include religious festivals, celebrations to mark important seasons, such as the start of planting or the end of harvest, and ceremonies for events within the life of a family or community, such as marriage, birth and death. An extension agent needs to know when these take place so that he can plan his activities around them. He should also take care to behave in the appropriate way on such occasions.

Traditional means of communication

All societies have ways of spreading information and sharing ideas. Songs, proverbs, drama, dancing, religious gatherings and village meetings are just a few of the traditional means of communication that an extension agent may find in a rural area.

There are two main reasons why these means of communication are important for extension:

- The extension agent can learn from them what people in the community are saying and thinking. An understanding of local proverbs, for example, will give the agent an insight into people's knowledge of their environment and their attitudes toward farming. Songs and dances often express deeply held feelings which an extension agent should be aware of when planning his programmes.
- The extension agent can make use of these traditional means of communication to pass on information and ideas. Many extension services now use drama, puppets and songs to convey new ideas.

Social and cultural change

Social structures and cultures are never completely static; they can and do change. The speed at which change takes place depends to a large extent on the contact people have with other cultures and new ideas, and on the ability of individuals within the society to initiate and accept change. Although the extension agent should respect and work through the existing culture and social structure, his task should be to help to speed up cultural change in farming. This may in turn contribute to wider social changes.

As ideas or methods are accepted within a society, they gradually come to be regarded as customary. A hundred and fifty years ago, land preparation in most of what is now Botswana was done with hoes. Farmers saw ploughs being used in what is now South Africa and introduced them to their own farms, with the result that an ox-drawn plough is now regarded as the normal equipment for land preparation and planting. More recently, in parts of Pakistan and Egypt, tractors are becoming part

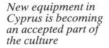

New equipment in Cyprus is becoming an accepted part of the culture

of the culture as they gradually replace draught animals as a source of power in farm operations.

New crops can also be introduced. Cocoa was unknown in Ghana until it was brought from the United States. Ghanaian farmers began to cultivate it in the nineteenth century when traders were keen to export the cocoa beans to Europe. Farmers learned the necessary techniques of raising young trees, fermenting and drying the beans and storage. Land-tenure rules changed as families moved to new areas to acquire land from other people on which to start cocoa farms. Cocoa gradually became a central part of Ghana's economy, tradition and culture.

As well as being aware of the social and cultural changes occurring in the area where he is working, the agent should try to understand the factors that can bring about such change.

Factors in change

Innovators

In every society, there are some individuals who are more ready than others to accept new ways of life. These people have a certain influence, but they can also often cause suspicion and jealousy among those who are less eager to change. However, if the new ways are seen to benefit those who have adopted them, the rest of the community may eventually come to accept them. The innovator may then be regarded without suspicion, and even gain in influence. General attitudes toward cultural change can then shift; new ideas may be welcomed as promising a better life instead of being regarded as a threat to established ways of doing things.

Contact with other cultures

Contact with other societies is an important force for cultural change. Cassava, for example, was first introduced to the west coast of Africa by Portuguese travellers who brought it from South America. It is now an important element in the diet in West African towns, and its introduction has led to many changes in farming systems. Similarly, maize spread from the United States throughout the world as people took it with them on their travels to other countries.

Extension agents often travel outside their areas in order to study. People who leave their society, to study or work among another society, bring back ideas which may change their way of life and be adopted by other people in their society. New styles of clothing, music, religious beliefs, house designs, political ideas and so on are spread from culture to culture by visitors and returning travellers. The more people are exposed to new ideas, the more likely it is that change may be accepted by the society as a whole.

Communication

Contact between different cultures is far more widespread than it used to be. New methods of communication bring societies throughout the world relatively easily into contact.

On a more local scale, roads and railways have brought many changes to rural society. Travel has been made easier and more people

Roads in Malawi bring shops, which introduce new goods and speed up cultural change

can visit other places and learn different ways of doing things. Traders establish shops and the goods in them may act as incentives for farmers to produce more in order to buy them. Crops can be marketed more easily and farming inputs brought into rural areas more quickly and cheaply.

Air travel has also had important effects. In Papua New Guinea, air services have enabled isolated mountain communities to market vegetables in towns and mining settlements that used to be inaccessible. The aeroplane has also helped to open up previously inaccessible areas of the Peruvian and Bolivian mountain regions. Villagers can now visit other communities and receive visitors from all over the world.

Newspapers, radio and television can also bring rural people in remote areas into contact with the outside world. People in rural communities who have radio sets or who read newspapers are usually influential and can spread their knowledge or new ideas to their neighbours. Education is another way of introducing people to the ideas, values and way of life of other societies.

Population growth

There is a close relationship between population size, farming systems and other aspects of culture. Where there are not many people in an area and there is plenty of farming land, farmers may abandon their fields after two or three seasons and move on to fresh, fertile land. The old fields then have a chance to recover during a fallow period. Whole villages may move as new land is cleared and prepared for farming but as population grows, land becomes scarce. New methods of farming have to be developed which allow fields to be cultivated year after year. Villages become permanent settlements. More elaborate houses can then be built because they do not have to be abandoned or moved every few years. As land becomes more and more scarce, individuals or families may move to other areas or to towns to look for work.

Economic factors

Economic development leads to changes in many aspects of people's lives and culture. The growth of towns and cities and the development of mines and industries have created new kinds of work in new places. People leave their rural homes to find work. In southern Africa, many men go to work in the mines and cities for a year at a time, leaving their wives to look after their farms. Jobs on the farm that were traditionally done by men now have to be done by women.

Elsewhere, on the fringes of the city, farming may become only a part-time occupation. Most families' main income may come from jobs in the city, but they keep their farmland as an insurance against unemployment and as a source of food. The presence of large numbers of part-time farmers will affect extension. Day-time meetings may be poorly attended, and part-time farmers will not necessarily be interested in new farming practices that increase output if it means spending more time working in the fields.

The growth of towns affects other aspects of culture, as well as the pattern of farming. Inheritance and land-tenure rules may change as people no longer have to rely on farmland to make a living. Where a lot of people from a village work in towns, they may be unable to attend traditional rural ceremonies and festivals which may then decline in importance. At the same time, those working in towns bring money and new possessions back to the village. These can improve rural living standards and have an important influence on values at the village level.

Social and cultural barriers to agricultural change

Although cultures and social structures are always changing, the process is often slow. In the short term, there will be features of society and culture that may act as barriers to change in agriculture. It is important that the extension agent be aware of the existence of such barriers and to take them into account in his work.

Respect for tradition

Many rural societies look upon new methods with indifference and sometimes with suspicion. Respect for elders often results in the attitude that the old ways are best. Farmers not only fear the unknown and untried but they also fear criticism for doing something different from other farmers. In such situations, the motives of extension agents and others seeking to promote change can often be misunderstood. Village people may think that the extension agent is introducing changes to benefit himself. Such attitudes explain the behaviour of farmers who seem to agree that a new method is good but are not prepared to put it into practice.

Belief in one's own culture

Members of all societies believe that their way of life is best. "These new methods of farming may be all right for some people but they are no good for us." This attitude results in reluctance to try something new. "How can it be better than our way?" and "We know what is best for us" are reactions that extension agents may meet in opposition to suggestions for change.

Pride and dignity

Farmers may be too proud to practise ways of farming that could result in other farmers looking down on them. For example, they might be too proud to carry cattle manure to the fields. Many young people leaving school look down on farming, even though some successful farmers earn more than most government employees and schoolteachers.

Relative values

Extension agents often emphasize the improved yield or cash return that can be gained by adopting new farm practices. However, farmers may value taste, appearance or some other factor more than the level of output. They may also value their leisure time so highly that they are not prepared to work longer hours on their farms. Certain improved varieties of

CHRISTOPHER M. WHITE

Cooking quality and taste may be more important to rural families than increases in output

maize have been rejected by small-scale farmers in several countries because of their poor flavour, even though they have shown a much better yield than local maize. Farmers and their families have to eat what they produce as well as sell some of it, so taste and cooking quality are very important.

Responsibilities and social obligations

Individuals within a society or a kinship group have responsibilities which they are expected to carry out. People who avoid such responsibilities anger other members of the society. As an individual's income increases, so obligations to society or family increase. The more money a farmer earns, the more help kinsmen will expect. This can be a very serious barrier to change if the individual sees little advantage in improving his or her position because there is not much personal benefit from the improvement. However, this may be overcome by concealing wealth, by distributing cattle among friends, or by burying or banking money so that relatives can be told that no money is available to help them. This may result, for example, in a farmer being reluctant to carry out visible farm improvements such as fencing, buying farm implements and other things which might suggest to kinsmen that the farmer is wealthier than they think.

Traditional ceremonies

Ceremonies such as weddings, funerals and religious festivals can take up so much time that the farmer may be unable to work to the maximum efficiency. The farmer is, therefore, unlikely to adopt new methods, which, while they might increase income, would mean that more time had to be devoted to working the farm and less to ceremonial and social obligations.

The extension agent needs to understand and to be sensitive to these potential social and cultural barriers to change; however, by carefully selecting what he encourages farmers to do, and how to convey the message, their effect can be reduced. Winning the support of traditional community leaders, for example, may lessen the effects of tradition. Furthermore, by making sure that popular food crops are included in agricultural programmes and that the recommended varieties are acceptable on grounds of taste and cooking quality, the extension agent can increase the likelihood of his advice being accepted. Extension programmes aiming at introducing new methods should take into account the possible effect on the whole society and its culture, and not merely the technical results of the methods recommended.

4. Extension and communication

Communication — the sharing of ideas and information — forms a large part of the extension agent's job. By passing on ideas, advice and information, he hopes to influence the decisions of farmers. He may also wish to encourage farmers to communicate with one another; the sharing of problems and ideas is an important stage in planning group or village activities. The agent must also be able to communicate with superior officers and research workers about the situation faced by farmers in his area.

There are many ways in which extension agents and farmers communicate. In this chapter, some general principles of communication will be looked at, and the use of mass media and audio-visual aids for communication in extension work will be discussed.

Communication

Any act of communication, be it a speech at a public meeting, a written report, a radio broadcast or a question from a farmer, includes four important elements:

- the *source,* or where the information or idea comes from;
- the *message,* which is the information or idea that is communicated;
- the *channel,* which is the way the message is transmitted;
- the *receiver,* who is the person for whom the message is intended.

Any communicator must consider all four elements carefully, as they all contribute to effectiveness. In considering each of these elements, the questions that follow provide a useful check-list.

Receiver

— What information does the receiver want or need?
— What information can he make use of?
— How much does the receiver already know about the particular topic?

CHRISTOPHER M. WHITE

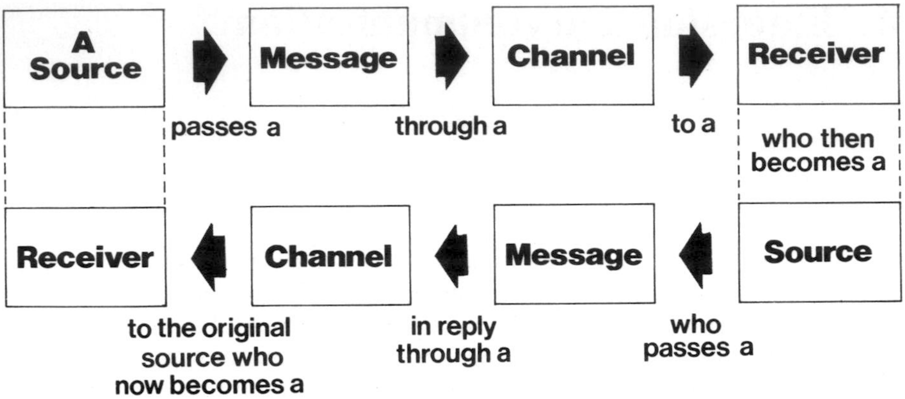

The four elements of communication

— What attitudes does the receiver hold concerning the topic?
— Should these attitudes be reinforced, or should an attempt be made to change them?

Channel

— What will be the most effective way of sharing the information? (This will depend upon the considerations outlined below.)
— What are the characteristics of the message? Does it need a visual presentation, as when crop pests are being described? Is it necessary to show movement or detailed actions (in which case, film, video or a demonstration will be needed)? If a permanent, accurate record of detailed information is required, as in farm records or fertilizer recommendations, the information should be in written or printed form.
— What channels are available to the receivers? Do they see newspapers? Can they read? Do many of them have radios?
— What are the receiver's expectations? A senior government official, for example, is more likely to take notice of a written submission followed by a personal visit.

Message

— What should the content be? A balance must be achieved between what the receiver wants to know and what the source feels the receiver ought to know.

— What form should the message take? In other words, how can the message be put into the words, pictures or symbols that the receiver will understand and take notice of?

Source

— Where will the information come from?
— Where should the information be seen to come from? An account of a successful cooperative in a nearby village may have much more effect if it is given by the members, through a radio programme or a visit, than by an extension agent at a public meeting.
— Has information from the source proved reliable in the past?
— How credible is the source in the eyes of the receiver?

Information often passes through several channels before it reaches a particular receiver, but it is rarely passed on in exactly the same words in which it was received. In particular, technical information is often distorted as it goes from one person to another. Extension agents should aim at being accurate sources and channels of information, and should make sure that farmers have heard and fully understood any information passed on to them. Leaflets and posters can be useful reminders of the spoken word.

Not all communication is deliberate. For example, people's behaviour, the way they speak to each other or the clothes they wear reveal much about them and their attitudes. If an extension agent is always late for meetings with farmers' groups, the members may come to the conclusion that he does not take them seriously. If he wears casual clothes when addressing a formal village meeting, villagers may say that he has no respect for them. Even if this is not so, the fact that they think it is will affect their relationship with the agent and, therefore, his effectiveness. The message that is received is not always the one that the source intends to pass.

Listening

A good communicator listens more than he speaks. An extension agent who does not listen to farmers and engage in a dialogue with them is unlikely to be very effective. There are four main reasons why a two-way exchange or dialogue is more effective than a monologue.

- Information needs can be assessed.
- Attitudes concerning the topic of the communication will emerge.
- Misunderstandings that occur during the exchange can quickly be identified and cleared up.

• Relationships of mutual respect can develop. If an agent listens, farmers will know that one agent is interested in them, and they will be more likely to pay attention to what the agent has to say.

Shared meanings

Communication is only successful when the receiver can interpret the information that the source has put into the message. An extension agent may give what he feels is a clear and concise talk, or an artist may be satisfied that he has designed a poster that conveys over the desired message, but there is no guarantee that those for whom the talk and poster are intended will interpret the message correctly. In the figure below, for example, the intended message is that crops should be rotated; however, many farmers may not understand the meaning of the arrows, or the symbols that stand for the different crops.

It is important that the same meanings for the words, pictures and symbols used in communication be used by the source and the receiver. If this does not happen, various kinds of problems can arise.

Language. Even if source and receiver speak the same language, local variations or dialects may use similar words with different meanings.

CHRISTOPHER M. WHITE

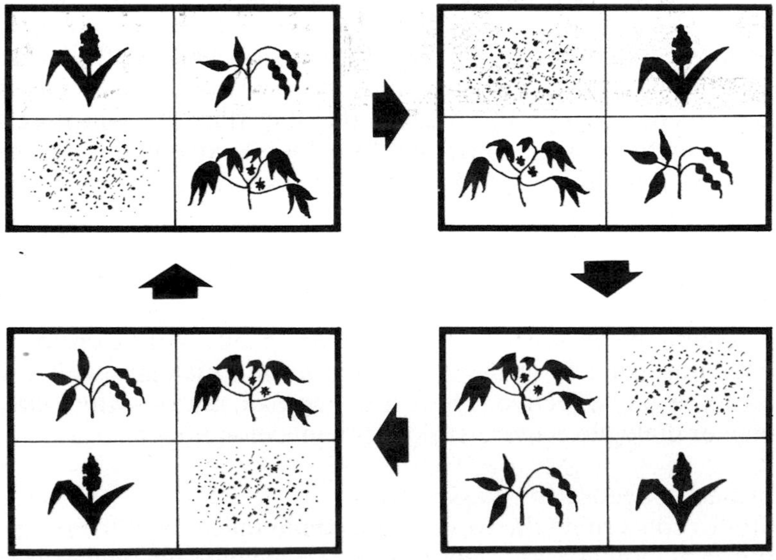

The agent may understand the message, but will the farmer?

Jargon. The technical language of specialists has to be translated into words that are familiar to the receiver. Extension agents need to learn what words and phrases farmers use when talking about their farming activities.

Pictures and symbols. Attempts to communicate through pictures and visual symbols often fail because the receiver does not recognize what they represent. Interpreting pictures is a skill which, like reading, has to be learned.

Mass media in extension

Mass media are those channels of communication which can expose large numbers of people to the same information at the same time. They include media which convey information by sound (radio, audio cassettes); moving pictures (television, film, video); and print (posters, newspapers, leaflets). The attraction of mass media to extension services is the high speed and low cost with which information can be communicated to people over a wide area. Although the cost of producing and transmitting a radio programme may seem high, when that cost is divided between the millions of people who may hear the programme, it is in fact a very cheap way of providing information. The cost of an hour's radio broadcast per farmer who listens can be less than one-hundredth of the cost of an hour's contact with an extension agent.

However, mass media cannot do all the jobs of an extension agent. They cannot offer personal advice and support, teach practical skills, or answer questions immediately. Their low cost suggests that they should be used for the tasks to which they are well suited. These include the following:

- Spreading awareness of new ideas and creating interest in farming innovations.
- Giving timely warnings about possible pest and disease outbreaks, and urgent advice on what action to take.
- Multiplying the impact of extension activities. A demonstration will only be attended by a small number of farmers, but the results will reach many more if they are reported in newspapers and on the radio.
- Sharing experiences with other individuals and communities. The success of a village in establishing a local tree plantation might stimulate other villages to do the same if it is broadcast over the radio. Farmers are also often interested in hearing about the problems of other farmers and how they have overcome them.

- Answering questions, and advising on problems common to a large number of farmers.
- Reinforcing or repeating information and advice. Information heard at a meeting or passed on by an extension agent can soon be forgotten. It will be remembered more easily if it is reinforced by mass media.
- Using a variety of sources that are credible to farmers. Instead of hearing advice from the extension agent only, through mass media farmers can be brought into contact with successful farmers from other areas, respected political figures and agricultural specialists.

Mass media communication requires specialist professional skills. Few extension agents will ever be required to produce radio programmes or to make films. However, extension agents can contribute to the successful use of mass media by providing material to media producers, in the form of newspaper stories, photographs, recorded interviews with farmers, items of information about extension activities or ideas for new extension films; and by using mass media in their extension work, for example, by distributing posters and leaflets or by encouraging farmers to listen to farm broadcasts.

Principles of media use

For extension through mass media to be effective, farmers must:

- have access to the medium;
- be exposed to the message: they may have radios, but do they listen to farm broadcasts?;
- pay attention to the message: information must be attractively presented and relevant to farmers' interests;
- understand the message.

Mass media messages are short-lived and the audience may pay attention for only a short time, particularly where the content is educational or instructional. If too much information is included, much of it will soon be forgotten. This means that information provided through mass media should be:

Simple and short.
Repeated, to increase understanding and help the audience to remember.
Structured, in a way that aids memory.
Coordinated with other media and with advice given by extension agents. It is important that what the farmers hear and see via mass media matches what extension agents tell them.

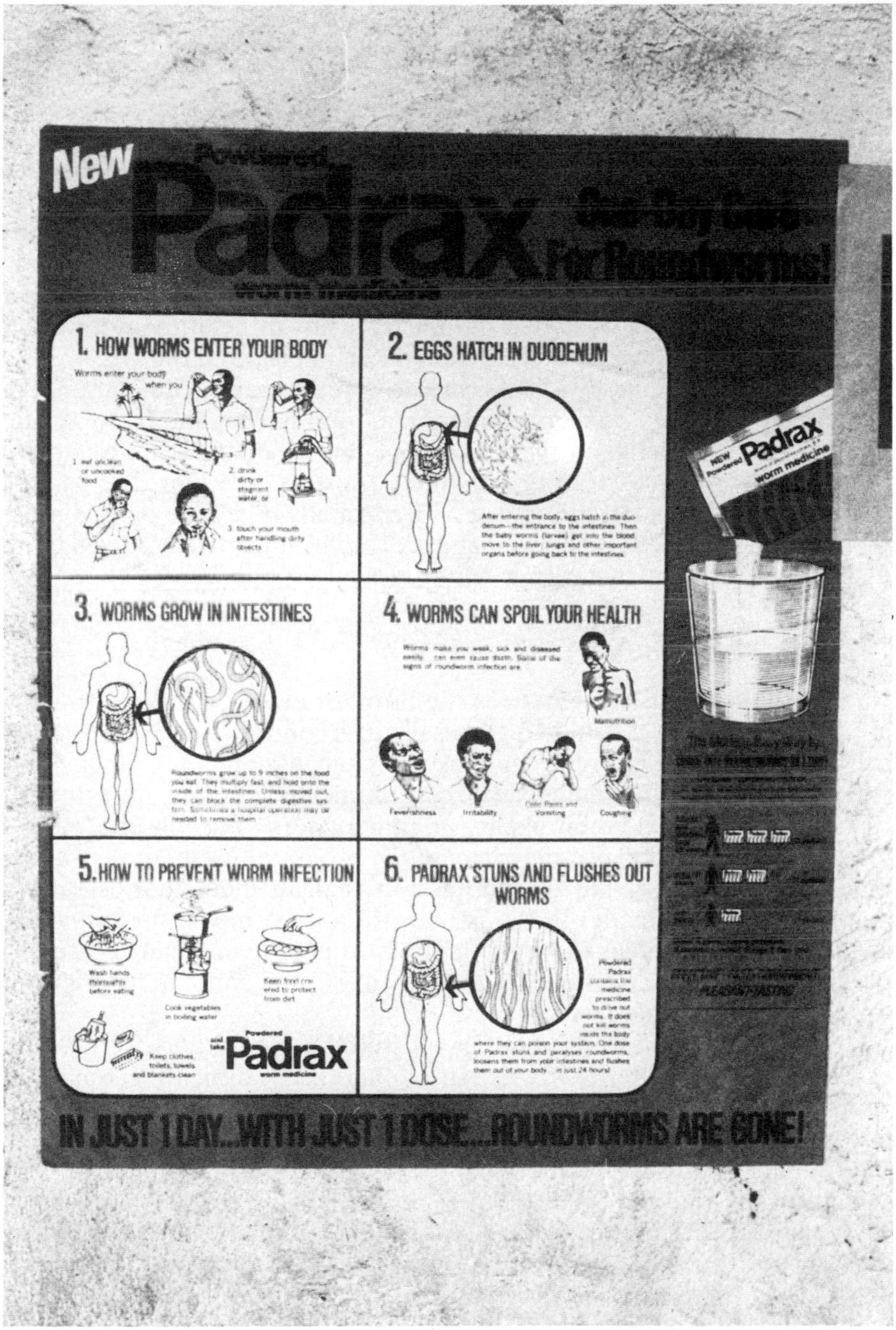

A poster on a shop wall in Malawi, containing several complex messages

Dialogue is also an important part of communication. With mass media, however, there is little opportunity for a genuine dialogue between farmers and those who produce the material. Consequently, media producers are not in a good position to determine farmers' precise information needs, or to check whether their messages are understood correctly.

One solution to this problem is for the producers to carry out research into farmers' existing knowledge, attitudes, practices, and problems concerning farming topics, and for mass media messages to be pretested. This means that a preliminary version of the message is given to a small number of farmers so that, if they have any difficulties interpreting it, revisions can be made before the final version is prepared.

Extension agents can help media producers by keeping them informed of farmers' concerns and information needs, and by reporting any failure to understand the content of the products of mass media. People who produce radio programmes, posters and films are usually more educated than farmers and are not normally in regular daily contact with rural people. They cannot, therefore, easily anticipate how well farmers will interpret the material they produce.

Radio

Radio is a particularly useful mass medium for extension. Battery-operated radios are now common features in rural communities. Information can reach households directly and instantly throughout a region or country. Urgent news or warnings can be communicated far more quickly than through posters, extension agents or newspapers. Yet, despite radio's mass audience, a good presenter can make programmes seem very informal and personal, giving the impression that an individual listener is being spoken to directly. Radio is one of the best media for spreading awareness of new ideas to large numbers of people and can be used to publicize extension activities. It can also enable one community or group to share its experiences with others.

There are, however, a number of limitations to the use of radio in extension work. Batteries are expensive and often difficult to obtain in rural areas, and there may be few repair facilities for radio sets that break down. From the listener's point of view, radio is an inflexible medium: a programme is transmitted at a specific time of day and if a farmer does not switch on the radio in time, there is no further opportunity to hear it. There is no record of the message. A farmer cannot stop the programme and go back to a point that was not quite understood or heard properly, and after the broadcast there is nothing to remind the farmer of the information heard.

A further limitation is the casual way in which people generally listen

Listening to the radio in groups encourages discussion, increases attention and promotes understanding

to the radio. They often listen while they are doing something else, such as eating, preparing food, or working in the field. For this reason, radio is not a good medium for putting over long, complex items of information. A popular format in many countries, therefore, is for short items of farming news and information to be presented between musical records. Radio drama, in which advice is given indirectly through a story or play, is also popular. This can hold attention and interest for longer than a single voice giving a formal talk. Finally, there is little feedback from the audience, except with a live broadcast where it is possible for listeners to telephone in their questions or points of view directly to the programme presenter.

Where there is only one national radio station, it may be difficult to design programmes that meet particular local needs. Moreover, it may not be possible to cater for variations in agricultural practices and recommendations in different areas. However, the growth in recent years of regional and local radio stations does make it possible for locally relevant information to be broadcast, and for extension agents to become more closely involved in making radio programmes. Local radio stations may be willing to allow extension agents to have a regular weekly programme; if so, they will usually offer some basic training in recording and broadcasting skills.

Farm broadcasts will only be attractive to farmers if they are topical and relevant to their farming problems. Extension agents can help to make them attractive by sending information and stories to the producers, and by inviting them to their area to interview farmers who have successfully improved their farms, or to report on demonstrations, shows and other extension activities.

Ways by which extension agents can achieve a more effective use of radio include:

Recording farming broadcasts on a cassette recorder for playing back to farmers later. This could greatly increase the number of farmers who hear the programmes.

Encouraging farmers to listen to broadcasts, either in their own homes or in groups. Radio farm forums have been set up in a number of countries; a group meets regularly, often with an extension agent, to listen to farm broadcasts. After each programme, they discuss the contents, answer each other's queries as best they can, and decide whether any action can be taken in response to the information they have heard.

Stimulating the habit of listening to farming broadcasts, and the expectation of gaining useful information from the radio. This can be done by the extension agent listening to the programmes and talking about the contents in his contacts with farmers.

Many extension agents will at some time have an opportunity to speak over the radio. They may be asked to interview farmers in their area or perhaps give a short talk themselves. The following guidelines for radio talks and interviews may be useful.

Radio talks

- Decide on the purpose of the talk; in other words, what you want people to know, learn or feel at the end of it.
- Attract attention in the first few seconds.
- Speak in everyday language, just as you would in a conversation, and not as though you are giving a lecture.
- Repeat the main points carefully to help the listeners to understand and remember.
- Give specific examples to illustrate your main points.
- Limit your talk to three minutes; the listeners will not concentrate on one voice speaking on a single topic for much longer than that.
- Make the talk practical by suggesting action that the listeners might take.
- Include a variety of topics and styles if you are given more than three minutes. A short talk could be followed by an interview or some item of farming news.

Interviews

- Discuss the topic, and the questions you intend to ask, with the interviewee beforehand.
- Relax the interviewee with a chat before beginning to record the interview.
- Avoid introducing questions or points that the interviewee is not expecting.
- Use a conversational style; the interview should sound like an informal discussion.
- Draw out the main points from the interviewee, and avoid speaking at length yourself; listeners are interested in the interviewee rather than you.
- Keep questions short; use questions beginning "Why"?, "What?", "How?" to avoid simple one-word answers, such as "Yes" or "No".

Audio cassettes

Audio cassettes are more flexible to use than radio, but as a mass medium they have their limitations. Cassette recorders are less common in rural areas than radio and are thus less familiar to villagers as sources of information. The cassette also has to be distributed physically, in contrast to the broadcast signal which makes radio such an instant medium. However, agents involved in many projects have found audio cassettes to be a useful extension tool, particularly where information is too specific to one area for it to be broadcast by radio.

The advantages of cassettes over radio are (*a*) that the tape can be stopped and replayed; (*b*) the listeners do not have to listen at a specific time of day; and (*c*) the same tape can be used over and over again, with new information being recorded and unwanted information being removed.

Information can be recorded on cassettes in a studio, where many copies can then be made for distribution, or it can be recorded on a blank cassette in the field. The possibility of recording farm radio programmes for playing back later has already been mentioned. Cassettes can also be used for:

Updating the extension agents' technical information. Pre-recorded cassettes, distributed by the extension organization, are a good way of keeping extension agents in touch with new technical developments in agriculture.

Sharing experiences between farmers' groups and between communities. An extension agent can record interviews and statements in one village and play them back in others.

Providing a commentary to accompany filmstrips and slide sets.
Stimulating discussion in farmers' groups or in training centres by presenting various points of view on a topic, or from a recorded drama.

Cassette recorders are light and fairly robust. However, they should be kept as free from dust as possible and the recording heads kept clean by using a suitable cleaning fluid, such as white spirit.

Film

The main advantage of film as a mass medium for extension is that it is visual; the audience can see as well as hear the information it contains. It is easier to hold an audience's attention when they have something to look at. It also makes it possible to explain things that are difficult to describe in words, for example, the colour and shape of an insect pest or the correct way to transplant seedlings. Moreover, by using close-up shots and slow motion, action can be shown in far greater detail than it is to see possible watching a live demonstration. Scenes from different places and times can be brought together in order to teach processes that cannot normally be seen directly. The causes of erosion, for example, can be demonstrated dramatically by showing how a hilltop stripped of trees no longer prevents rain-water running down the slope, creating gullies and removing topsoil. Similarly, the benefits of regular weeding can be shown by filming crops in two contrasting fields at different stages of growth. Once a film has been made, many copies can be produced with the result that thousands can then watch the film at the same time.

Films come in two formats: 16 mm and 8 mm. Most cinema and educational films are in the larger 16-mm format. Equipment and production costs for 8-mm films are much lower, but because the picture quality is not quite so good and the projected picture size is relatively small, 8 mm has until recently been regarded as suitable for amateur domestic use only. As equipment improves, however, more organizations are producing training and educational films in 8-mm format. An 8-mm film cannot be shown on a projector made for 16-mm films or vice versa. Whichever format of film is to be used, it is necessary to have a projector; a screen or a white wall on which to project the film; a loudspeaker for the film's soundtrack (unless it has no soundtrack, in which case the extension agent may need a microphone, amplifier and loudspeaker so that he can give his own commentary); and a power source, which will either be mains electricity or a generator. If a generator is used, it should be as far away as possible from the projector and the audience so that its noise does not distract them from the film.

Because films require this cumbersome equipment, it is not practical for the extension agent to show them in villages unless he has motor trans-

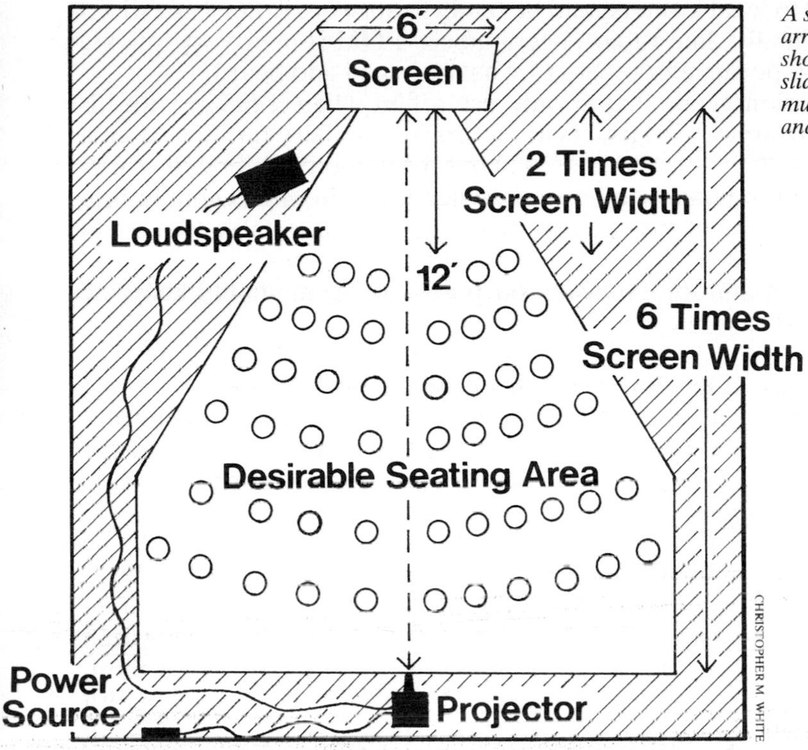

A suitable arrangement for showing films or slides: the audience must be able to see and hear clearly

port. It is more common for mobile cinema vans to bring films to rural areas, or for films to be shown in schools and rural training centres where equipment is available.

Film has a number of other limitations as a medium for rural extension. A film may take several months to produce since filming, processing, editing and copying all take time. Films are also expensive to make, and are worth making only if they can be shown many times over a number of years. They are, therefore not a good medium for topical information which soon becomes out of date.

The films seen by rural audiences have often been made in areas that are very different from those where they are shown. It may be difficult for the audience to relate their own farming to the crops, livestock, farm implements, people and housing that they see on the screen. The content may therefore seem of little relevance to them. Furthermore, there is no opportunity for a dialogue between film producer and farmer. Extension agents should, therefore, preview a film wherever possible, be prepared to explain the relevance of the information it contains whenever the details may be unfamiliar to local farmers, and be ready to answer farmers' questions afterwards. Finally, like radio programmes, a film is over very quickly and there is no permanent record of what was seen and heard.

An extension agent should only use a film when it fits in with his extension programme. If farmers are interested in dairy farming, then a film on the topic can give some ideas about the equipment, breeds of cattle and forms of organization they might need. Again, if an agent wishes to spread awareness of the dangers of soil erosion, a suitable film could explain the causes and effects as well as control measures.

When using film for extension purposes, an agent should keep the following points in mind.

Select films which fit in with the extension programme.

Publicize the film, after selecting a suitable date and venue in consultation with local leaders. Films are best shown in the evening; if the weather is suitable, the film can be projected against the outside white wall of a school or other building.

View the film in advance, and decide if the information needs to be adjusted to suit local conditions. This can be done either by speaking to the audience afterwards, or by turning the sound commentary off and giving a verbal explanation while the film is being shown.

Try out the equipment, especially if there is to be no technician present. It is useful to know how to change the bulb in the projector, for example, as these occasionally break.

Follow up the film by discussion and questions to help the audience to understand the content, relate it to their own situation and remember it.

Television and video

Television, like film, combines vision with sound and like radio, it can also be an instant medium, transmitting information directly to a mass audience. Television signals can be broadcast from a land-based transmitter, by satellite or through cables. However, in many countries, television transmission and sets are still restricted to urban areas, and the potential of television for rural extension will remain low until sets become more widely available. Television sets are much more expensive to buy and repair than radios, and programme production costs are also far higher. Where television has been used for rural extension communication, access and impact have been increased by group viewing followed by discussion.

Video combines most of the advantages of film and of audio cassettes. Using a video camera, picture and sound are recorded on a magnetic tape and are then immediately available for viewing on a monitor or television set. This enables the production team to re-record any material that is not satisfactory. As with audio cassettes, unwanted information can be removed and the tape reused.

As a mass medium, video has more to offer than film, since video programmes can be made far more quickly in multiple copies, and the lightweight video cassettes are relatively easy to distribute. As video equipment — television monitors and video cassette recorders — becomes more robust, it will be possible to use mobile units to show up-to-date programmes, made within the country and even within the area, to large numbers of rural families. The tape can be slowed down, wound back to repeat a particular action, or held on a particular frame while the extension agent explains a point. The same mobile units could carry portable video cameras to collect material for new programmes. The main limitation to viewing is that only 20 to 30 people can satisfactorily watch a video programme on a normal television set, while several hundred can see a film projected on to a large screen.

Where video equipment is available — and it will become increasingly so over the next few years — extension agents should refer to the guidelines given above for using film and audio cassettes.

Printed media

Printed media can combine words, pictures and diagrams to convey accurate and clear information. Their great advantage is that they can be looked at for as long as the viewer wishes, and can be referred to again and again. This makes them ideal as permanent reminders of extension messages. However, they are only useful in areas where a reasonable proportion of the population can read.

Printed media used in extension include posters, leaflets, circular letters, newspapers and magazines.

Posters are useful for publicizing forthcoming events and for reinforcing messages that farmers receive through other media. They should be displayed in prominent places where a lot of people regularly pass by. The most effective posters carry a simple message, catch people's attention and are easy to interpret.

Leaflets can summarize the main points of a talk or demonstration, or provide detailed information that would not be remembered simply by hearing it, such as fertilizer application rates or names of seed varieties.

Circular letters are used to publicize local extension activities, to give timely information on local farm problems and to summarize results of demonstrations so that the many farmers who cannot attend them may still benefit..

F. MATTIOLI

A poster in India carrying a simple message, and located beside the village water supply to attract maximum attention

Newspapers are not widely available in rural areas. However, local leaders often read newspapers, and a regular column on agricultural topics is useful to create awareness of new ideas and to inform people of what other groups or communities are doing.

Printed media can be either very sophisticated, with colour photographs and a variety of lettering styles, requiring expensive equipment that is only available in large cities, or produced simply and cheaply using equipment found in many local extension offices, such as a typewriter, stencils, a duplicator and a photocopier. This simpler technology makes it possible for extension agents to produce leaflets and circular letters that are relevant to their area and can be made available quickly to farmers. With the use of two duplicators — one with black and one with red ink — quite attractive leaflets can be produced. Stencil duplicators cannot reproduce photographs, so illustrations must be limited to simple outline drawings and diagrams. Modern photocopiers, however, can produce reasonable copies of black-and-white photographs.

Where the extension agent is using printed material that has been mass produced, he should make sure that it complements his extension activities. Posters may be used, for example, to draw attention to a topic related to a later demonstration, but printed material that the farmer does not see as relevant to what the extension agent does or says will have little impact.

Printed media are of little use if they are not distributed. Expensively produced posters, leaflets and magazines should not be allowed to gather dust on extension office shelves: they should be made widely available and farmers should be encouraged to look at and discuss them. Posters should be replaced regularly with new ones. In addition, where printed material proves to be irrelevant or difficult for farmers to understand, those who produced them ought to be informed so that improvements can be made. Posters and leaflets that seem clear to the extension agent may not be fully understood by farmers. Whenever possible, the agent should help to explain their meaning. In time, farmers will become used to the ways in which pictures and words convey information and will find it increasingly easy to interpret printed media.

When the agent is preparing his own printed media, or material is being produced to his specifications, the following stages offer a very useful guide. They apply equally to posters, leaflets, circular letters and newspaper articles.

Define the context. The agent should be clear about the purpose of the material. Is it intended to create awareness and stimulate people to seek more detailed information? Or to remind farmers of what they have learned? Or to provide detailed technical information and serve as a reference for future use? The agent also needs to know how the material will be used by the audience. Will it be seen casually as people pass by a notice-board? Will it be studied individually in the home, or discussed at a group meeting?

Know the audience. Before planning the content, the agent needs information about the particular audience: their knowledge and attitudes concerning the subject-matter of the information, and their farming practices.

Decide on content. The information must be relevant to farmers' needs, and the content and amount of information should also suit the context in which the media will be used. A poster, for example, should contain one simple message in large, readable type that can be interpreted quickly by a passer-by.

Attract attention. The material must be attractive at first glance. Only if a person's attention is caught by a leaflet or a poster will he spend the necessary time to look at, read and absorb the information it contains. This can be helped by short, boldly printed headings, eye-catching pictures and sufficient empty space to prevent it from looking too dense or cluttered.

Structure the information. The agent can help farmers to understand and remember the information by dividing the contents into sections that lead logically from one to another, and by the use of headings and underlining to bring out the main points.

Pre-test. All locally produced material should be pre-tested before use. It can be shown to a few people from the target group, who should then be asked what information they have learned from it. This gives an opportunity to improve the material, if necessary, before beginning final production.

Exhibits and displays

Apart from being a useful way of sharing information, an attractive, neat display suggests to people that the extension agent and his organization

KENYA INFORMATION SERVICE

A mobile extension exhibit in Kenya uses real objects and demonstrations to help people visualize the physical benefits of increased milk consumption

are efficient and keen to communicate. Displays are suitable for notice boards inside and outside extension offices, at demonstration plots (where the progress of the demonstration can be recorded in pictures), and at agricultural shows. Although a good display can take quite a long time to prepare, it will be seen by many people. With displays on permanent notice-boards, it is important that the material be changed regularly so that people develop the habit of looking there for up-to-date information.

A display should stick to a single theme broken down into a small number of messages. It should include several pictures (preferably photographs) and diagrams which must be clearly labelled. If there is a lot of printed text that is not broken up by pictures, the display will look dull and fail to attract attention.

Campaigns

In an extension campaign, several media are used in a coordinated way and over a limited period of time in order to achieve a particular extension objective. The advantage of campaigns is that the media can support and reinforce one another. The disadvantage is that campaigns can take a lot of time and effort to plan. Often the extension agent will be involved in campaigns planned by staff at national or regional level. His role will be to make local arrangements for meetings, film shows, demonstrations advance publicity, accommodation for visiting staff and distribution of printed material.

An extension agent can also plan his own local campaigns. A campaign can be useful in situations where the farmers of an area face a common problem for which there is a solution which could readily be adopted. Campaigns require careful planning to make the best use of all extension methods and media available. Principles of extension planning (see Chapter 7) and guidelines for the various methods and media should be used in planning campaigns.

Traditional media

Traditional forms of entertainment can also be used as extension media. Songs, dances and plays can convey information in an interesting way. Even when they are prepared in advance, they can be adapted at the last minute to cater to local situations and response from the audience. No modern technology is required and these media are especially useful where literacy levels are low. By involving local people in preparing the plot of a play, extension agents can stimulate the process of problem analysis, which is a fundamental part of the educational aspect of extension.

Audio-visual aids in extension

The term audio-visual aid refers to anything that an extension agent uses to help to convey the message when communicating with farmers. The spoken word is the agent's main communication tool, but, whether the agent is speaking to a large village meeting or discussing a problem in a field with a group of farmers, its impact and effectiveness can be greatly increased by the use of suitable audio-visual aids. When selected and used properly, audio-visual aids can help in the following ways:

- The interest of the audience can be maintained if the agent varies the mode of presentation. It is difficult to concentrate for long on what someone is saying; but if the agent refers to a wall chart, or illustrates a point with some slides, his audience's attention can be maintained.
- When information is presented to more than one sense (sight and touch, for example, as well as hearing), more is taken in and it is better understood and remembered.
- Processes and concepts that are difficult to express in words alone can be explained. The procedure for applying for a loan, for example, may sound confusing, but a simple chart or diagram can make the process clearer. Again, the life cycle of a crop pest can be explained by showing a series of slides or drawings.
- The effects of decisions and actions that farmers might take can be shown. Photographs of a cattle dip or a model of a cooperative store can give farmers a clear idea of just what it is they might be considering.
- Pictures can have a more immediate impact on our emotions than words. Photographs of a heavy crop, for example, are likely to arouse interest more effectively than details of yields read out by an extension agent.

The range of audio-visual aids

Extension agents often use sophisticated audio-visual aids which require electricity and complex machinery such as projectors or television sets. But there are many simple aids that the agent can make locally, and these have several advantages. They do not require a power source or heavy equipment, they do not cost much to produce and they can be made to suit the precise needs of the agent. Between these two extremes lies a wide range of more or less sophisticated aids. The distinction between a mass medium and an audio-visual aid lies only in the way it is used. All the mass media described earlier can be used as audio-visual aids. A film is a mass medium, in that it is shown to large audiences in many different places; but for an individual extension agent who uses it to increase the impact of a talk, it is an audio-visual aid. Many of the principles of media use discussed earlier also apply to audio-visual aids. The audio-visual aids available to the extension agent can now be examined.

Objects

A real object is often the most effective aid. It enables the audience to understand exactly what the extension agent is talking about. Equipment and tools can be shown, samples of diseased plants and insect pests displayed and different seed varieties and fertilizers handled by farmers.

Where an object is too large to be shown, a model of it can sometimes be used as a teaching aid. This applies particularly to buildings and other fixed structures. The construction of a poultry shed, for example, or the installation of a dip tank can be demonstrated by using a model, which can be taken to pieces in front of the audience.

Photographs offer another substitute for real objects. They can be passed around an audience or displayed by the agent. If a photograph is

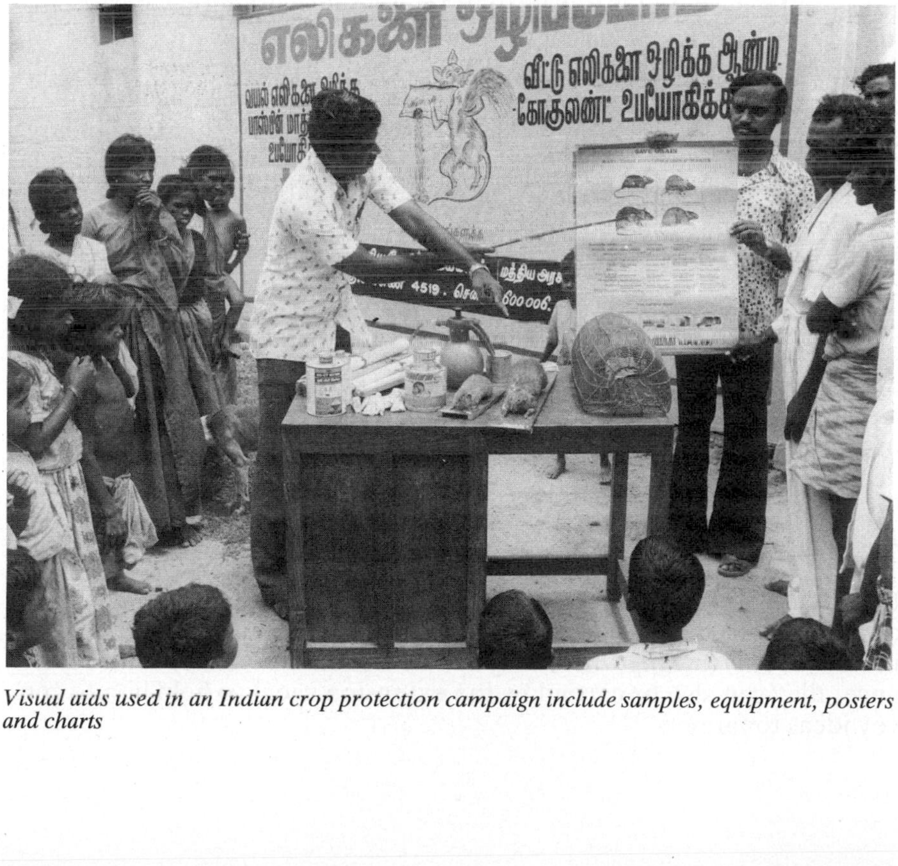

F. MATTIOLI

Visual aids used in an Indian crop protection campaign include samples, equipment, posters and charts

being taken to use as a visual aid, just the right amount of detail should be included for the audience to recognize it. Too much detail confuses and distracts, while too little prevents recognition. Photographs of people doing things are more likely to interest the audience than photographs of objects alone.

Chalkboards

Blackboards are widely available in schools, rural training institutes and extension offices. They may be fixed to an inside wall or supported on a free-standing easel which can be moved around. They are useful for setting down the main headings of a talk, for sketching simple drawings and diagrams, and for noting points raised in questions and discussion.

If using a blackboard, the agent should practise writing on it, if necessary by drawing horizontal chalk lines for guidance. He should make sure that the writing is large enough for someone at the back of the audience to see clearly and that the headings and phrases are kept short. There is not much space on blackboards and the agent will lose the audience's attention if he spends a lot of time with his back to them while writing.

Whiteboards have a smooth, shiny surface on which coloured felt pens can be used, but it is important to use only pens with water-soluble ink. Whiteboards are easier to use than blackboards from both the agent's and the audience's point of view. The pens flow smoothly over the surface and the colours are much clearer than chalk on a blackboard.

Newsprint, which is an inexpensive paper, can be obtained in large sheets and fixed to a blackboard or to the walls of a building. It can be used in the same way as a blackboard but it is more versatile. Text and drawings can be prepared on several sheets, before a meeting, to avoid having to write while speaking. Paper and pens can be given to small discussion groups to note their conclusions. These conclusions can then be displayed around the meeting-place and discussed by others. Suggestions and ideas from the audience can be added to enable farmers to see their decisions taking shape. Used sheets can be kept for future reference. At a planning meeting with a group of farmers, for example, the agent can take away the sheets to guide him in preparing a written record of the decisions taken.

Posters

Posters are useful for highlighting the main theme of a talk and wall charts can be used to show complex processes. Although they are used mainly in class-room teaching where they can be left on the wall for future reference, they can also be carried by the extension agent to help him to convey ideas to farmers.

An extension agent pre-testing a flip chart in Botswana

Flip charts

Flip charts contain a series of pictures, with or without words, fastened along one edge between two sheets of thin wood or thick cardboard. The two covers can be opened and folded back so that the flip chart stands in front of the audience. Each picture illustrates one point in the extension agent's talk and he simply turns over each one when he moves on to the next point. As well as helping the audience to understand and remember, they remind the agent of the structure of his talk without the need to refer constantly to his written notes.

Many extension agents will already have access to printed posters, wall charts and flip charts, but they can also be made locally with large sheets of paper and coloured pens. When making flip charts, the following points should be noted.

- Lettering should be large.
- Diagrams should be simple.
- Information on each sheet should be limited.
- Pictures from posters and magazines can be cut out and stuck on by those who cannot draw.
- Pre-testing is important for all home-made visual aids.

Flannelgraphs

A flannelgraph is made from rough textured cloth, such as flannel or a blanket, which is hung or supported almost vertically. Figures, words, and symbols cut from cardboard, which are backed with similar cloth or sandpaper, are attached to it. A cheaper backing is obtained by putting glue on the back of the cut-out and then dipping it into fine sand. The backing holds the cut-outs firmly on the cloth surface. The cut-outs are prepared beforehand and can be used repeatedly.

The flannelgraph can be used very effectively to build up a story or an explanation. Unlike a wall chart, which can confuse an audience by presenting a finished diagram at the start of a talk, a flannelgraph can be used to present in turn each part of the diagram until it is complete.

The cut-outs can be placed in different positions to show alternative outcomes. After showing the process of wind erosion, for example, the effect of wind-breaks can be demonstrated by placing cut-outs of trees between the wind direction and a field. Arrows representing the wind can then be deflected, and the general effect shown by putting back soil symbols on the surface of the field.

A modern alternative to flannelgraphs is the magnetic board. Cut-outs are backed by a magnetic strip, that holds them firmly to a metal board. They can be used in windy conditions when flannelgraph cut-outs would blow away, but they are cumbersome to transport. On the other hand, flannelgraphs, which can be made in a variety of sizes and designs can be folded into an agent's bag or rolled up and tied to a bicycle.

Projected aids

Films, colour slides, filmstrips and overhead projector transparencies are useful as teaching aids, bringing colour, variety and interest to an extension talk. However, they all require specific equipment and electricity. Extension agents are, therefore, more likely to use them in training centres and schools, although some slide projectors can be adapted to work from a 12-volt car battery. Films, filmstrips and slides are best used at night or in a room with curtains drawn or shutters closed. Daylight screens can be used for small groups. Overhead projectors can be used in daylight, provided the sun is not shining directly on the screen or wall on which the image is projected.

Colour slides can be selected and put in a suitable sequence by the extension agent. He can produce his own slides to suit his purposes, provided he has access to a camera, film and film processing facilities. A slide set can easily be modified or updated by replacing one or more slides. If they are kept dry and free from dust and fingerprints, they will remain in good condition for many years. An agent can either provide his own spoken comments on the slides, or a commentary can be recorded on an audio cassette. With synchronized equipment, the tape can be modified so that slides automatically change at the appropriate point.

Filmstrips contain a sequence of slides in a single continuous strip of film. They are shown on a slide-projector fitted with a filmstrip carrier between the projector body and the lens. They cannot be modified easily and the sequence is fixed, but individual frames cannot fall out or be put into the projector the wrong way round. They are useful when a fixed message has to be presented many times.

Overhead projectors are usually only found in class-rooms. Diagrams and texts are put on to a sheet of transparent acetate with special felt pens; the acetate is then placed on a flat glass platform through which a light shines, projecting the contents on to a vertical screen. The agent can write on the acetate while facing his audience, or he can prepare it beforehand. If he covers different parts of a sheet with paper, he can gradually reveal the sections of a diagram, thus achieving an effect similar to the flannelgraph.

Using audio-visual aids

Audio-visual aids are only effective if they are appropriate to the situation and are used properly by the agent. Unsuitable aids or ones that are not used properly can at best distract and at worst mislead the audience. When selecting suitable audio-visual aids, the agent will be limited to what is readily available or can be made. Within that range, some aids are more suited to a particular objective than others. For example, if accurate detail is needed, a photograph, slides or a careful drawing may be more appropriate. If, on the other hand, the agent simply wants to highlight the structure of a talk or the main conclusions of a discussion, a blackboard or newsprint will be suitable. The agent should also consider where the aids will be used: indoors or outdoors, with or without electricity, at a large meeting or with a small group. All these factors will influence the choice of audio-visual aids.

Proficiency in using audio-visual aids cannot be learned from a book; it comes only with practice. The following principles may, however, be useful, whatever audio-visual aids an extension agent may use.

Select the aids most in accordance with your objective, the composition and size of the audience where the aids will be used.

Use the aids to reinforce your message. They are there for support, to complement and supplement the spoken word, and should not be expected to communicate their contents without explanation. Refer to them, explain them and ask questions about them.

Make sure that the audience will be able to see and hear clearly. Audio cassettes that cannot be heard or lettering that is too small to be seen can make the audience restless and inattentive.

Practise using the aids beforehand. Where projected aids are used, it is important to be completely accustomed to the equipment. For example, there are seven incorrect ways of loading a slide into a projector but only one correct way.

5. Extension methods

In the previous chapter, the mass communication methods that the extension agent can employ in his work with farmers were reviewed. In this chapter, two other extension methods that the agent can employ will be examined. They are (*a*) the individual method, in which the agent deals with farmers on a one-to-one basis; and (*b*) the group method, in which the agent brings the farmers together in one form or another in order to undertake his extension work. Each of these methods demands different approaches and techniques on the part of the agent, and these will be examined later.

The two methods are suited to different purposes. It is important for the extension agent to consider the range of individual and group methods at his disposal and to select the method appropriate to the situation. It is also important to remember the educational purpose of extension work, and to ensure that the method selected is used to promote the

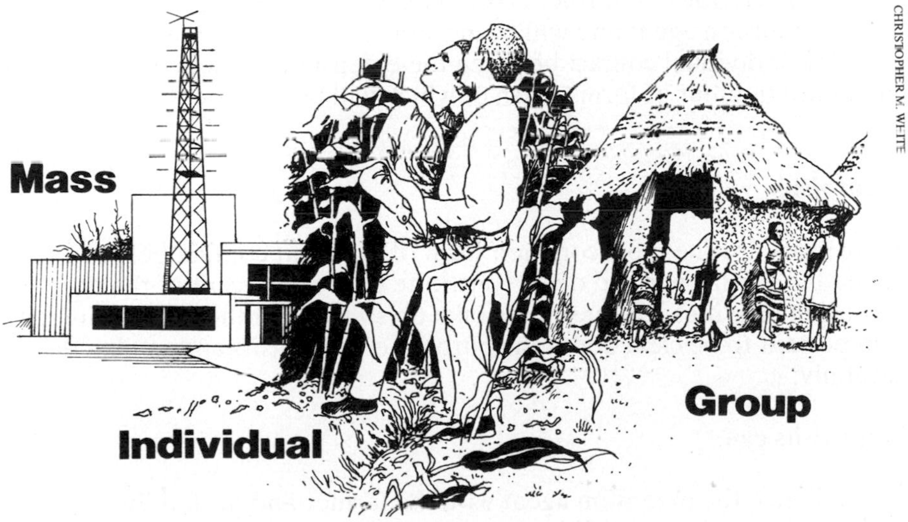

Three different extension methods: mass, group and individual

farmers' better understanding of the technology involved. Both individual and group extension methods involve the agent in a face-to-face relationship with the farmer, and this relationship should be one of mutual confidence and respect. The agent, therefore, must think carefully about his use of individual or group extension methods and ensure that his relationship with the farmer is sensitively developed.

Individual methods of extension

Individual or face-to-face methods are probably the most universally used extension methods in both developed and developing countries. The extension agent meets the farmer at home or on the farm and discusses issues of mutual interest, giving the farmer both information and advice. The atmosphere of the meeting is usually informal and relaxed, and the farmer is able to benefit from the agent's individual attention. Individual meetings are probably the most important aspect of all extension work and invaluable for building confidence between the agent and the farmer.

Learning is very much an individual process and, although group methods enable the agent to reach a greater number of farmers, personal contact with and the individual attention of the extension agent are important supports for a farmer. The personal influence of the extension worker can be a critical factor in helping a farmer through difficult decisions, and can also be instrumental in getting the farmer to participate in extension activities. A farmer is often likely to listen to the advice given by the extension agent and will be grateful for this individual attention.

This individual contact between the extension agent and the farmer can take a number of forms, each of which will be considered below.

Farm visits

Farm visits are the most common form of personal contact between the agent and the farmer and often constitute over 50 percent of the agent's extension activities. Because they take up so much of the agent's time, it is important to be clear about the purpose of such visits and to plan them carefully.

Farm visits can:

- familiarize the extension agent with the farmer and his family;
- enable him to give specific advice or information to the farmer;

A coffee extension agent in Colombia visits a farmer on his coffee holding

• build up the agent's knowledge of the area, and of the kinds of problems which farmers face;
• permit him to explain a new recommended practice or follow up and observe results to date;
• arouse general interest among the farmers and stimulate their involvement in extension activities.

At times, the extension agent will make a farm visit spontaneously if he happens to be passing by and it is convenient to drop in. Such informal visits may have no specific purpose but are a useful way of maintaining contact and gradually building links with farmers. Even if the agent just drops in to greet the farmer and his family, this short visit can do a lot to foster mutual respect and friendship. Usually, however, farm visits will be part of the agent's general plan of work and will be programmed into his monthly schedule of activities.

Planning the visit

First, it is important to be very clear about the purpose of the visit. Before a visit, the extension agent should review the file on the farm to be visited and consult any other information available on the farmer. He must be fully informed on the relevant details of the farm he is visiting and should summarize the situation in a few notes before setting off. He must try to avoid showing ignorance of the farmer and his farming activities, or the need to consult his file during the visit.

In addition, the visit should be planned to fit in with other local extension activities. For example, if a demonstration or meeting is planned for the morning, then it may be possible to programme a number of individual farm visits for the afternoon. Whenever possible, the agent should make an appointment at a time convenient to the farmer, to ensure that the farmer will be there and that the journey will not be wasted. If an appointment is made in advance, the farmer will also have time to prepare for the visit and to think about the issues to be discussed with the agent.

Check-list

- Make an appointment if possible.
- Decide the purpose of the visit.
- Review previous records and information.
- Prepare specialist subject matter that might be required.
- Schedule the visit into the overall work plan.

Making the visit

The agent must always remember the basic educational purpose of extension and his role in this process. The agent's role is not just one of transmitting new knowledge or recommendations; he must also devote time during visits to building up the farmer's confidence and interest. One of the first points to think about when making a farm visit is how to start a conversation. The first few minutes of contact are extremely important for establishing a good relationship, particularly if it is the first meeting. Methods of establishing rapport and of initiating conversation differ from culture to culture. Small talk in order to break the ice is often an important first step, and gives both sides a chance to relax and to get to know each other a little before more serious matters are discussed. Time must be spent in greeting the farmer and his family and an informal chat will not be wasted. Similarly, local customs should be followed as regards accepting hospitality. If visitors are expected to drink tea or coffee with the host, then the agent should do so, while taking care not to acquire a reputation of one who spends all his time drinking tea or coffee during visits.

The agent should then choose the moment when more formal business can be discussed. The choice of the discussion topic is also an important decision for the extension worker. If he is sensitive to the farmer's needs, he will discuss matters relevant to these needs. Moreover, he will discuss these topics at the farmer's level and in language that the farmer is accustomed to use. In this respect, the agent should be a good listener as well as a good talker, and he should encourage farmers to explain and discuss issues at their own pace and in their own words. It is important for the agent to find a reason to praise farmers for some aspect of farm management in order to encourage their involvement and make then feel that they also have knowledge to contribute.

The visit may cover a whole range of activities. The farmer may need further explanation or information about a particular new practice and it may be necessary to demonstrate this. If further technical information is required, this can be noted and, at a later date, an appropriate information bulletin can be sent. The agent may also brief the farmer generally on government agricultural policy, or describe the range of extension activities in the area and encourage the farmer to get involved. Some visits may also be of an emergency nature when the agent will be expected to give on-the-spot advice concerning a particular problem. Finally, the agent should always be alert during farm visits for ways to promote the involvement of other family members, including spouses and young people, in other local extension activities such as a youth club or a food preparation demonstration.

If possible, it is useful to set up a system to keep a record of the content of farm visits; this could be simply a notebook in which the essential details of each visit (date, purpose, problems, decisions, etc.) are recorded. Such a farm record system is very useful in helping the agent to keep a more detailed account of each visit and it is invaluable when a new agent is taking over. The records are also very useful in evaluating farmers' progress.

Check-list

- Be punctual for the visit.
- Greet the farmer and this family.
- Praise the farmer's work.
- Encourage the farmer to explain and discuss any problems.
- Provide any technical or other information required.
- Record the details of the visit.
- Plan with the farmer the time and purpose of the next meeting.

Recording and follow-up

The purpose of any farm visit will be lost if its content and conclusion are not recorded and no follow-up action taken. During the visit, the agent will almost certainly make a series of notes and observations. On the first available office day, these notes should be neatly transcribed on an individual card kept at the extension office. After each visit, the agent should record the date, the purpose, the conclusions or recommendations arising from the visit, as well as any other additional information or observations which he feels will be helpful. The agent must not store this information in his head, since, if he is transferred, the incoming agent will be left with a gap in information and will find it difficult to catch up with the situation.

Finally, the agent should arrange for any follow-up that may have been agreed with the farmer. This could involve the sending of further specialist information, or arranging for a specialist colleague to visit the farmer concerned. Whatever the case, it is vitally important that the agent follow up on any issues or problems that he was not able to deal with in person. Failure to do so will disappoint the farmer and lessen his confidence in the agent. It is very important to maintain the confidence and trust that can often take years to build up. He should also schedule his next visit to the farmer in his work programme.

Check-list

Record purpose of and decision resulting from visit.
Arrange for any follow-up information or advice to be sent.
Schedule the next visit to the farmer.

Farm visits are perhaps the single most important aspect of the agent's work in terms of establishing rapport with the farmers in his area and of building the trust and confidence that are vital to the success of his work. However, farm visits take up a lot of time and only a few farmers can be reached. Farm visits are, therefore, a costly extension method and for this reason they must be carefully thought out and planned. The visits must make an impact and must lead to positive agricultural development if they are to justify their cost. Finally, the agent should beware of visiting some farmers repeatedly. This would not only severely limit the range of his activities, but could also arouse the resentment of other farmers who might feel excluded.

Office calls

Just as the extension agent visits the farmer, so he can expect that from time to time the farmer will visit him at his office. Such a visit is often a reflection of the interest which the agent may have aroused among the local farmers. The more confidence local farmers have in the extension agent, the more likely they are to visit him. Such office visits are less time-consuming for the extension worker, and offer some of the advantages of a farm visit. While no extension agent would wish to be overwhelmed by such visits every day, he should encourage farmers to drop in if it is convenient for them to make the exchange of visits two-way.

As with farm visits, office visits similarly have to be prepared. Although the agent may not know when a farmer is likely to drop in, he can at least arrange the extension office in such a way that the visitor feels at ease and can understand the activities of the office. The arrangement could include:

- ensuring that access to the extension office is adequately posted and the agent's name displayed;
- having a notice-board clearly displayed upon which useful, up-to-date information can be pinned;
- having one or two chairs where visitors can wait for appointments;
- displaying any bulletins, circulars or other written extension literature that the visitor can read.

For some farmers, a visit to the extension office may be a difficult experience. The agent should, therefore, try to put the farmer at ease, asking a few questions in order to get the farmer to explain his problems. The agent should be polite but purposeful, and try to find out the reason for the visit as quickly as possible. When he feels that the matter has been adequately discussed, the agent should tactfully terminate the interview

CHRISTOPHER M. WHITE

The layout of an extension office is important

in order not to let it drift aimlessly on. The agent should always escort the visitor out, and say goodbye. A note on these office visits should also be added to the farmer's record card, and any follow-up implemented.

Letters

Occasionally, the extension agent will correspond with a farmer by letter. Letters can be a follow-up inquiry resulting from an agent's farm visit, or sent because a farmer is unable to make a personal office visit. Drafting and replying to letters are very important skills for the extension worker and he should give every thought to them. Problems can arise with the use of words or complex technological language, or if the letter has been badly typed or written. In writing a letter to a farmer, the extension agent should try to put himself in the farmer's shoes. The letter should be in the local language, preferably not on impressively headed writing-paper, and should always contain some personal greeting to the farmer. Often, farmers will show such letters to their neighbours and thus it is important to create a favourable impression. The following points are important:

- letters should be clear and concise, so as not to confuse the reader;
- the information in the letter should be complete and relevant to the issues raised;
- where possible, letters should be answered promptly. If time is needed to collect information for the reply, a short letter of acknowledgement should be sent;
- a copy of the letter must always be made and entered in the office file.

Other individual methods

Telephone calls

Telephone calls and office visits serve a very similar purpose. It is improbable that the extension worker will deal with many of the farmers in his area (if at all) by telephone. If the telephone is used, however, it will not be for long discussion but for passing on specific advice or information. Whatever the reason, it is important for the agent to speak clearly, to note the main points discussed and to enter them on the farmer's record.

Informal contacts

Informal contacts will occur continually during the agent's stay in a particular area. Market days, holiday celebrations or religious events will bring him into contact with the farmers with whom he is working who will inevitably talk about their problems. By attending such events, the agent can become well acquainted with the area where he works and with the farmers and their problems, and he will be able to pass on ideas and information on an informal basis.

Group methods of extension

The extension agent should consider the use of the group approach in his work with farmers. The use of groups in extension has become more common over the past decade, and indeed a number of new ideas have emerged about how groups may be used most effectively. For example, the widespread Small Farmer Development Programme (SFDP) in Southeast Asia was based upon group methods and it has produced two manuals which detail the approach of group extension work. Furthermore, in Latin America, work with extension groups in Brazil and Colombia has shown the usefulness to extension of the formation of extension groups, and how these groups can support extension activity.

Advantages of group methods

It has been seen that individual extension methods can be costly in both terms of time and scarce extension resources, and that they reach only a limited number of people. There is also the danger that too much emphasis upon individuals can lead to undue concentration on progressive farmers to the detriment of the poorer farmers.

Coverage

The group method offers the possibility of greater extension coverage, and is therefore more cost-effective. Using the group method, the extension worker can reach more farmers and in this way make contact with many more farmers who have had no previous contact with extension activities.

Learning environment

Extension groups offer a more reflective learning environment in which the farmer can listen, discuss and decide upon his involvement in the extension activity. The support of the group helps an individual farmer to

Group of smallholders discussing their problems with an extension agent in Peru

make decisions and determine a course of action. The group creates a supportive atmosphere, and individual farmers can gain greater self-confidence by joining others to discuss new ideas and try out new practices.

Action

The group method brings together farmers with similar problems. Often, these problems demand concerted action (tackling the erosion of a hillside, for example), and such action can be taken more effectively by a group rather than by an individual, who may be overwhelmed by the enormity of certain problems.

Important issues in group extension

Before considering in detail a number of different group methods of extension, it is necessary to look at some of the more important issues concerning the extension group. To form, structure and develop a group of farmers for extension purposes is a complex process, and such groups do not appear overnight. It is not sufficient for the extension agent merely to bring the farmers together for a particular activity. He must give time and thought to the fact that the farmers will constitute a group, will function as a group and will display characteristics associated with groups. Experience in different parts of the world has shown that there are four sets of important issues that the agent will have to bear in mind:

Purpose

The agent should be aware of two main purposes in his work with groups. First, he should try to develop the group, to encourage its members to continue to meet and to establish the group on a permanent basis. In this way, the agent will be developing a base from which group members can continue their development efforts. Second, the agent should use the group to transmit new ideas, information and knowledge that will assist the farmers in their agricultural activities. While the second use of groups is more common, it is important that the agent consider the initial development of the group as an equally vital extension activity.

Size

The most suitable size for groups in rural extension is between 20 and 40 members. If the group is too large, it becomes unwieldy and many farmers may feel lost and bewildered. Smaller groups allow closer contact, a better chance of involvement and more opportunity for strengthening bonds of friendship and support among members. One common determinant of group size is geographical location: its membership will be restricted to those living within a particular area.

Membership

Since the extension agent's job is to help farmers identify and tackle problems, it is better to have groups of farmers with common problems. If the agent is working with a group made up of different types of farmers, ranging, for example, from big landowners to small, tenant farmers, it may be difficult for him to achieve a common purpose within the group. The agent, therefore, should pay careful attention to group membership and try to ensure that its members share a common interest and problems.

Agent's relationship with group

The agent should give considerable thought to his relationship with the group. Ideally, he will want to encourage the group's formation and help to strengthen it. If his extension work takes him to another area, it is hoped that he will leave behind a structure that can function with a minimum of extension support. The agent should try to avoid being directly responsible for setting up the groups and should try to ensure that they are based, where possible, upon existing social or cultural community groups. In all of his activities with groups, the agent should beware of the group becoming too dependent upon him and of creating a structure that needs him for survival. Instead, the agent should strive to encourage an element of independence in the group, by encouraging the group to take the initiative in extension activities and to decide for itself in what way the agent can be of assistance.

These four issues, then, should be borne in mind by the agent as he pursues his work with extension groups. His main concern will be to do his work well and to ensure that, through the group approach, more farmers come into contact with new ideas and practices. In the long run, group work might be even more productive and effective if he gave some thought to the development of the group itself in his extension work.

Types of group extension methods

Group meetings

Calling the members of a group or the inhabitants of a local community together for a meeting is the commonest group extension method. Although there may be an air of informality about such meetings, they will nevertheless need to be carefully thought out and planned. The group or community meeting is a useful educational forum where the agent and farmers can come together, and ideas can be openly discussed and analysed. The agent will probably have information about a new government policy, or agricultural idea or practice. He will want to introduce this new information, to seek the opinions of community members and gain their support for extension activities. Indeed, there are a whole range of purposes for such community or group meetings:

Information meetings. The agent calls the group or community together to communicate a specific piece of new information which he feels will benefit them and upon which he seeks their advice.

Planning meetings. The main purpose is to review a particular problem, suggest a number of solutions and decide upon a course of action.

Special interest meetings. Topics of specific interest to a particular group of people (e.g., horticulture, bee keeping, or dairy farming) are presented and discussed in detail at a level relevant to those participating.

General community meetings. Men, women and young people of a community are invited to attend to discuss issues of general community interest. It is important to hold such general meetings occasionally so as to avoid any community group feeling that it is excluded from extension activities.

Whatever the case, however, the agent should only call a meeting if he thinks that it can be useful. If farmers feel that their time has been wasted in coming to a meeting, they may refuse to come to subsequent meetings and thus frustrate the agent's work. Once he decides to hold a meeting, the agent should make careful preparations and check a number of important arrangements which will be necessary to ensure a successful meeting.

The basic purpose of the meeting should be agreed and to determine this the agent should consult community or group leaders. Only then can the agent and community leaders consider the content and the best approach to the meeting. It may be useful to write down in a few words what the purpose is, and then to see what are the important aspects to be considered. If it is to be a meeting for providing information, the agent must structure his material in a coherent form and decide in what sequence he is going to present it. If it is to be a general community meeting, then, similarly, he must decide how he will structure the meeting and introduce discussion on the issues he has in mind.

Form of the meeting

Depending upon the nature and purpose of the meeting, the agent must decide the most appropriate form for the meeting and how it can best be conducted. The agent should consider the appropriateness of the different forms the meeting could take and, in consultation with community members, decide accordingly. Ideally, the ingredients should be mixed to suit the occasion.

Smaller meetings are more likely to meet the specific needs of those who attend. When plans are to be made or decisions taken, a small number of representatives will usually achieve more than a large gathering of all community members. On other occasions, it will be important for the meeting to be open to as many people as possible.

A formal meeting, with chairman, agenda and written record of proceedings, is appropriate when specific business has to be dealt with or decisions reached. The chairman keeps the meeting to the central issues, and the decisions of the meeting are recorded accurately so that they cannot be disputed later. In an informal meeting, people feel more able to express their own point of view and less dominated by the structure and formality of the proceedings. However, an open, unstructured discussion, although it allows all to participate, may result in a few people dominating the proceedings.

A lecture or talk allows the agent (or other speaker) to convey a detailed, well-prepared message to his audience on a specific issue; for example, a new piece of technology can be presented in this way, and illustrated by visual aids. It should be remembered, however, that the lecture is a particularly tedious approach to meetings and care must be taken to ensure that people will not get bored. Alternatively, in a discussion, many people are able to express points of view and ask questions. Discussions may be completely open and unstructured, or based on a prepared agenda of discussion points.

Planning the meeting

There are two important decisions to make regarding the time and location. A date and time for the meeting must be decided and announced. The time should be convenient to all concerned and should avoid clashes with other events or activities. The meeting-place should be well-known, easy to get to and appropriate for the form of meeting. The meeting-place should also be comfortable and have the facilities necessary for the meeting. An extension agent would never hold a meeting at midday, on a very hot day, in the open sun. Such a meeting could be disastrous, as well as cause considerable discomfort.

After the above two issues have been considered, it may be useful for the agent to draw up a list of other arrangements to be made in preparation for the meeting. Such a check-list could include:

Check-list

- Publicity for the meeting
- Seating arrangements
- Audio-visual equipment and material, or other educational aids
- Agenda, and order of events
- Guest speakers or other specialists who will contribute to the meeting
- Chairman to take charge of the meeting, who should be elected by the community
- Refreshments for speakers and, where necessary, other participants.

F. BOTTS

A forestry extension agent meets a local group in Nepal

Conducting the meeting

Even the most carefully prepared meeting can fail if it is not conducted in the right manner. While the above arrangements are important, the way the actual meeting proceeds will determine whether it will be a success or not. The agent must be conscious that he is dealing with adults who do not want to sit for hours listening to a speaker talk endlessly. The agent should try to vary the agenda of the meeting: for example, a short talk, accompanied by visual aids, followed by comments and questions.

Variety of content, as well as a chance for the farmers to participate, will be important. In addition, the meeting must not go on too long. One-and-a-half hours are probably sufficient for a group or community meeting. It is better to have a highly productive, short meeting than one which rambles on and loses effect.

The agent's role in the meeting should also respond to the circumstances. He should encourage the community to appoint a chairman and should allow the chairman to conduct business. The agent's role should basically be to inform and support, and he should not dominate the meeting. Furthermore, the meeting should not resemble a class-room with the agent as teacher and the farmers as pupils. The agent should make every effort to ensure that during the proceedings the community members feel that it is their meeting and that they have a part to play.

As a guide to the proceedings of the meeting, the agent should keep the following points in mind. He should start the meeting on time. Then he should welcome community members and special guests, explain the purpose of the meeting and the programme to be followed, and begin the programme. Later, the agent should encourage questions and discussion, and be prepared to summarize the main points and note important decisions. The meeting should be closed with thanks to all concerned.

As the most commonly used form of group extension method, the group or community meeting will be most effective if carefully thought out and planned. After each meeting, the agent should make a brief record of the proceedings and the principal decisions taken. He should also take any prompt follow-up action that has been decided.

Demonstrations

Farmers like to see how a new idea works, and also what effect it can have on increasing their crop production. Both purposes can be achieved by means of a farm demonstration. A good, practical demonstration is an invaluable method in extension work. The demonstration is a particularly powerful method to use with farmers who do not read easily. A demonstration will give such farmers the opportunity to observe, at first hand, the differences between a recommended new crop practice and traditional practices. The strength of the demonstration should lie in its simplicity and its ability to present the farmers with concrete results.

There are two principal types of demonstration used by extension agents — method demonstration and result demonstration.

Method demonstration

Method demonstrations basically show farmers how to do something. In the method demonstration, the farmer is shown step by step how, for example, to plant seeds in line, to use a mechanical duster to control insects, or to top tobacco. The agent will probably be dealing with farmers who have already accepted the particular practice being demonstrated, but who now want to know how to do it themselves.

The main advantage of the method demonstration is that the extension agent can explain simple farming skills to a large number of people, thus increasing the impact of his extension work. Moreover, as farmers are able to participate, there is a greater chance that they will benefit from the demonstration than if they were passively hearing it in a lecture.

F. MATTIOLI

Conducting a method demonstration in food preparation in El Contentó, Colombia

The main limitation of a method demonstration is that, if there are too many farmers present, only a few get a chance to see, hear and do. The agent must be conscious that the demonstration is a learning experience and prepare the event accordingly. It is also vital that the demonstration be well thought out and competently conducted.

Result demonstration

The main purpose of a result demonstration is to show local farmers that a particular new recommendation is practicable under local conditions. Comparison is the important element in a result demonstration: comparison between compost and no compost, between poor seed and selected seed, or between use of fertilizer and no fertilizer. "Seeing is believing" is an age-old expression, but one appropriate to a result demonstration. Until a farmer has actually seen the results of, for example, the application of a fertilizer, he will not be convinced by the agent's recommendation. By showing tangible results of a new practice recommended by the extension service, the agent can help to create confidence among the farmers and can greatly encourage them to try the practice themselves.

F. BOTTS

Result demonstration: comparing rice grown with and without applications of fertilizer in Jogjakarta Province, Indonesia

A result demonstration is an ideal way to present to farmers a comparison between traditional and new practices. It can also help to establish confidence in more scientific farming methods and increase the farmers' confidence in ideas originating from research stations. It shows proof of the value of a new practice. A result demonstration is also a useful tool that an agent can use to establish confidence among farmers in a new area.

Its major limitation is that it takes a long time to mature and is thus a costly use of extension resources. If, in the end, for whatever reason, the new practice should fail, it could have disastrous consequences. Often such failures (for example, because of lack of rain) are outside the control of the agent.

Both method and result demonstrations are extension activities that require a lot of thought, careful planning and efficient execution. Although the two demonstrations differ somewhat in their purposes, they share a lot of common points and, in terms of their preparation and execution, they can be considered together.

Basic principles for demonstrations

Before the agent begins to plan and prepare for a demonstration, he should be clear about a number of key points that will guide his preparation and handling of the demonstration.

Participation. Where possible, demonstrations should be carried out on local farms with farmers' participation rather than on an extension plot or research station. Farmers will have more confidence if a demonstration is held on a neighbour's land, or if a new practice is shown by a fellow farmer, than if it is carried out by agents on extension land. The more the local farmers can be involved in the whole process of a demonstration, the greater will be their self-confidence and readiness to learn.

Simplicity. Simple, clear-cut demonstrations of a single practice or new idea will be far more effective than ambitious and over-complex demonstrations that demand too much of the farmer. It is better to proceed step by step with a number of demonstrations than to try do to everything at once.

Learning. The demonstration is a learning environment and should be run in such a way that the farmers do in fact learn something. A demonstration is a type of class-room, and the agent must be conscious of class-room requirements in terms of space, time, equipment and the teaching method to use.

Preparation. An extension agent should never contemplate holding a demonstration without careful planning and preparation. A demonstration hastily given could have disastrous consequences.

Planning the demonstration

When the agent decides that a demonstration would be useful at a particular time, he must then dedicate some time to planning and preparing for it. In this respect, he must ask himself a number of questions.

— What is the objective of the demonstration?
— Why is the demonstration the most suitable extension method, and what would be the usefulness of the new idea to be demonstrated?
— When should the demonstration be held? When is the most convenient date and time both for the farmers and in terms of the application of the new idea?
— Where is the demonstration to be held? Which suitable location is the most convenient for the farmers.

The agent should work out in some detail his answers to the above questions before proceeding any further. It is very important that the reasons for the demonstration be appropriate and clearly understood and that there is a realistic expectation that the demonstration will be of benefit to the farmers involved.

Preparing the demonstration

The more carefully the agent can prepare all the details of the demonstration, the more chance he will have of it running smoothly. The following are the key areas of preparation.

- Consult the local people and seek their help and advice in the preparation of the demonstration.
- Prepare a detailed plan of the demonstration, the main issues to be covered, the sequence of events, the resources needed and the contributions required from other people.
- Collect information and material available on the new idea or practice to be demonstrated, and make sure that the topic is familiar and that questions can be answered.
- Check that all the support material is ready (e.g., audio-visual aids, implements).
- Select those farmers who will take part in the demonstration and brief them on the outline of events.
- Ensure that the demonstration has been publicized and that the farmers know exactly when and where it is to take place.
- Visit the demonstration site beforehand to make sure that all is in order and that the site is appropriate.

Supervising the demonstration

During the demonstration, the agent's role should be to supervise but not to dominate. He should actively support the farmer who may be assisting in the demonstration, and encourage the others to participate as much as they can. The agent should be keen to ensure that all those present benefit from the demonstration. During the demonstration, therefore, the agent should:

Welcome the participants, make them feel at ease and ensure that they have all they require to benefit fully from the demonstration.

Explain the purpose of the demonstration, what it is hoped to achieve and what the various stages are that will be followed. Distribute any literature or other material which may have been prepared as a guide for the participants.

Conduct the demonstration in person or be ready to help the demonstrator farmer. Proceed at a pace the farmers can follow, and be prepared to explain again or answer questions from participants. Emphasize key points and explain the practice step by step in simple words. In a method demonstration, ensure that all those who wish to do so have a chance to practise the demonstration themselves.

Summarize the main issues or points which have arisen, encourage questions from the farmers and make sure that the participants have had every opportunity to try out or examine the practice being demonstrated.

Conclude the demonstration with a vote of thanks to all concerned, and with a few comments about any follow-up activities planned.

Follow-up

It is important that any interest generated by, or decisions taken at, the demonstration be followed up. Farmers will feel let down if the agent does not do so. This follow-up will be useful for the agent as well. Demonstrations can often result in good contacts with local farmers, and the agent may be able to enlist their support for future activities. It is also important that the agent reflect upon the demonstration and evaluate its effectiveness. The agent should, therefore, write a report and prepare a record of the demonstration, noting the names of the participants, the effect achieved and personal impressions of the usefulness of the demonstration.

Field days

Field days are usually opportunities to hold method or result demonstrations on a slightly larger scale, and are usually run in a more informal and less highly structured manner. The purpose is often to introduce a new idea and a new crop, and to stimulate the interest of as many farmers as possible. Experimental stations or other government centres may be used for field days, but it is more usual and profitable for them to be held on the land of a local farmer. There is a greater chance of making an impact if the field day is held on a farmer's land, and if the farmer plays a part in running it and explaining the purpose.

Field days can range in size from a small group to annual events attracting hundreds of farmers. Since the aim is a general introduction to some new idea, there is less need to be concerned about limiting the numbers. The extension agent's role on the field day is to support the farmer on whose land it is being held, to offer general guidance to ensure that things run smoothly and to be available to answer questions and queries.

It is probably better not to over organize the field day but to try to create an atmosphere in which visiting farmers can inspect, inquire, question and generally get to know what is available.

Although the agent will try to encourage an open and informal atmosphere for the field day, there is still a considerable amount of preparation needed to ensure that it runs well. The issues which the agent must consider are very similar to those noted under demonstrations and will not be repeated here. It may be useful, however, to bear a few additional points in mind.

Limit the numbers to the capacity of the field, to avoid overcrowding.

Ensure a good layout of field-day activities, with easy access and facility of movement around the field.

Encourage the demonstrator farmer to take most of the initiative; give him support but do not take over the field day from him.

Provide suitably large visual material and also, if necessary, a loudspeaker, to ensure that all can hear. Check that extension literature and other material are available for consultation.

Conclude the field day by bringing all the participants together, reviewing the day's proceedings and the main items seen and discussed, and explain any future relevant extension activities.

A field day is a day out for farmers and is often a welcome relief from their daily hard work. The agent should, therefore, provide an interesting and well-presented exhibition, suitable refreshments and points of rest, and generally create an atmosphere in which the farmers will feel at ease and will be eager to know what is going on.

Tours

Farmers like to visit farms in other districts to see how they work, what they grow and what kinds of problems the farmers there are facing. A tour is a series of field demonstrations on different farms, or at different centres, and can often attract a lot of interest from local farmers. The tour should give local farmers a chance to see how other farmers cultivate their land, and to exchange ideas and experiences with them. It is important, therefore, that the area to be visited be in some way similar agriculturally to that of the visiting farmers.

As with all other forms of extension, tours have to be well thought out, planned, prepared and conducted. The five stages of determining the objective, planning the content, preparing the arrangements, conducting the tour and arranging for appropriate follow-up will be a guide to the extension agent. However, it may be useful to add these points:

Visit the area first to become familiar with local conditions, the farms to be visited, the route and road conditions.
Limit the tour to what is possible. It is better to do a short tour in which visitors can have a good look at local farms than to arrange an ambitious tour and be pushed for time. Don't tire the visitors out.
Encourage the host farmers to do all the explaining and to take charge of the tour.
Arrange for food and drink during the tour.
Conclude the tour with a short summary of the main events and note any comments or conclusions.

A field tour is an ideal method of involving farmers and of stimulating genuine interest in extension activities. It is also very useful in bringing farmers together to discuss common problems, and to gain useful experience of other areas.

6. The extension agent

The whole extension process is dependent upon the extension agent, who is the critical element in all extension activities. If the extension agent is not able to respond to a given situation and function effectively, it does not matter how imaginative the extension approach is or how impressive the supply of inputs and resources for extension work. Indeed, the effectiveness of the extension agent can often determine the success or failure of an extension programme.

The extension agent has to work with people in a variety of different ways. It is often an intimate relationship and one which demands much tact and resourcefulness. The agent inevitably works with people whose circumstances are different from his own. He is an educated, trained professional working with farmers, many of whom have little formal education and lead a way of life which may be quite different from his.

Extension agent

In his extension work, the agent basically intervenes in the life of the farmers in a particular area. The extension agent is a change agent: he intervenes to bring about change in order to help improve the lives of the farmers and their families. This is not an easy task and a series of issues arise in relation to this intervention. The basic role of the agent in bringing change into a rural area and what areas of knowledge and personal skills would be useful in performing this role both need to be considered in this case.

The role of the agent

There are no models of an agent's role which are applicable to all situations. An agent must consider each situation individually and adopt a position or role suitable to that situation.

Indeed, there is a wide variety of views on the extension agent's role in bringing about change among farmers. To illustrate this range of views, a number of different statements on the agent's role, taken from extension practice from different parts of the world, can be examined.

- An extension agent tries to arouse people to recognize and take an interest in their problems, to overcome these problems, to teach them how to do so, to persuade them to act on his teaching, so that they ultimately achieve a sense of satisfaction and pride in their achievements.
- A change agent is a person whose primary role is to achieve a transformation of attitudes, behaviour and social organization.
- Change agents are multi-purpose agents serving as links between government and people.
- A change agent is a person who sets in motion a process of change after realizing that certain changes are necessary for the rural society.
- A change agent is an activist whose main role is to help people form their own organizations in order to be able to tackle their problems.
- A change agent is a professional who influences the innovation/decision-making process in a direction deemed desirable by the change agency.

The above statements capture the flavour of the wide-ranging views that exist on the role of the extension or change agent. It is not suggested that any one statement is more valid or important than the rest. They are merely presented to show the complexities of the agent's role and to stress the need for very careful thought by the agent before embarking upon a process of change.

However, a lot of the ideas about the agent's role can be assembled into two very broad categories. The agent can be seen as having two main, but different, areas of responsibility.

Knowledge/communication/innovation

The extension agent is responsible for providing the knowledge and information that will enable a farmer to understand and make a decision about a particular innovation, and then for communicating that knowledge to the farmer. In this role, the agent is seen as a vehicle of knowledge, usually of a technical nature, and as a teacher who instructs farmers in the use of this knowledge. The agent is formally trained for this position and is provided with the technical knowledge and information which he must then communicate to the farmers. In this role, the agent's work is usually highly structured and based on existing government policies and programmes of rural development.

Educator/facilitator/catalyst

In the role of educator, facilitator or catalyst, which the agent may need to perform in the course of his duties, the agent is associated less with the knowledge/communication aspect and more with the farmers' personal development. The agent is less concerned with specific programmes or targets and more with helping the farmers to gain confidence, to organize themselves and to begin to get involved in extension activities. The agent's role is essentially to help support and actively encourage farmers to develop their own initiatives and to begin to tackle their own problems.

This discussion can be summarized by listing the wide range of key words used in extension throughout the world to describe the role of the extension agent. The list is not intended to suggest that the agent must be all of these things. It does, however, underline the agent's importance in extension practice throughout the world, and the many-faceted interpretation of this role:

Teacher	Facilitator	Organizer	Arbitrator
Educator	Broker	Administrator	Advocate
Leader	Consultant	Enabler	Catalyst
Communicator	Intermediary	Activist	Friend
Motivator	Listener	Provider	Stimulator

The above list shows us the diversity of roles an agent can assume, but he must always be aware that the most important thing is to study the

I. VELEZ

An extension worker in Honduras helping a rural women's cooperative to increase their incomes with apiculture

situation, analyse the problems and adopt a position which is relevant to solving those particular problems. Thought must be given to this, and an agent must never simply plunge into a situation without thinking carefully how he may best help to change it.

Knowledge and personal skills

Two other important issues are the types of knowledge that an agent must have, and the personal skills required of him to do his job effectively. Again, in extension practice throughout the world, there is considerable diversity on these two issues, reflecting the variety of situations in which agents work. How the knowledge and personal skills required by the agent are influenced by the role the agent performs will be considered.

Knowledge

Four main areas of knowledge are important for the extension agent and form the basis of extension training.

Technical. The agent must be adequately trained in the technical aspects of his work and have a good working knowledge of the main elements of the agricultural system in which he is working.

Rural life. This includes anthropological and sociological studies of the rural area where he is working, local traditions, practices, culture and values.

Policy. The agent should be familiar with the main legislation of government or other institutional policies which affect the rural areas, development programmes, credit programmes, and bureaucratic and administrative procedures.

Adult education. Since extension is an educational process, the agent must be familiar with the main approaches to adult education and group dynamics, and with the techniques of developing farmer participation in extension activities.

It should be noted that these four areas of knowledge can be dealt with by a training programme, when the agent can be provided with the detailed information available in the four areas.

Personal skills

It is more difficult to determine the personal skills required of the agent for extension work, and to train the agent in these skills. The area of personal skills is less easily covered by means of specific knowledge and information and often refers to skills that an agent either has or has not. A vast range of such skills has been suggested; however, these have been grouped together to present a list of the main areas of skills required of an extension agent.

Organization and planning. The extension agent must be able to plan extension work, to organize its implementation and generally to manage and effectively control an extension office and its activities.

Communication. An extension agent must above all be a communicator, both verbally and non-verbally, and this skill is the basis of all extension activity.

Analysis and diagnosis. The extension agent must be able to examine situations which confront him, recognize and understand the problems that exist and propose courses of action.

Leadership. The extension agent should inspire confidence and trust in the farmers he serves, set them an example and take the lead in initiating activities.

Initiative. The extension agent may often have to work in isolation and unsupervised. He must have the initiative and confidence to do so without depending upon guidance and support from his superiors.

Personal qualities

The personal qualities required by a good extension worker are often discussed. These qualities are more difficult to define but, nevertheless, they are qualities to look for when selecting extension agents. Some of the qualities suggested are personal characteristics, and it is important to assess whether an agent possesses them before appointing him to an extension post. The personal qualities required in an extension agent include:

- Commitment to extension work and to working, at times, in isolated rural areas, with a sense of dedication and determination to get some extension activities under way.

Extension agents

• Reliability, both in terms of carrying out extension work and also in relations with farmers. An extension worker's superior officers must be able to rely on him to carry out his tasks without close supervision, and the local farmers must come to have confidence in his advice and support.
• Humility in his work with the farmers. The agent must be sensitive to the wishes and feelings of the farmers and work with them in a way that respects them as people who have knowledge and ideas to contribute.
• Confidence in his own abilities and determination to achieve something. An extension agent is often left to work in isolation with little supervision and needs self-confidence and courage to do so.

The above areas of knowledge, personal skills and qualities are not exhaustive. They are not presented as a check-list against which to judge the competence of an individual to do extension work, but to show the very demanding nature of the work and to act as a guide when selecting and training extension agents.

Public speaking

Public speaking is a skill which the agent will be called upon to practise frequently. A principal task of the agent is communication; this inevitably involves public speaking to explain a new idea, conduct a demonstration or generally take part in a community discussion.

Speaking in public is a very useful and effective form of communication, if done competently. A public speech gives the extension agent an opportunity to demonstrate his enthusiasm and technical knowledge. Some people are natural public speakers and easily and competently arrange their material and present it in a way which is both interesting and intelligible to the audience. On the other hand, if an extension agent is poorly prepared to give his speech and delivers it badly in an uncomfortable situation, then his efforts could be counterproductive.

Public speaking represents an important aspect of an extension agent's work and one which he can use to great advantage if he prepares well. Most importantly, a talk must flow well and give the impression of being well-thought-out. There is nothing more off-putting for an audience than a speaker who fumbles his notes, forgets basic facts or talks too long. A well-prepared and delivered extension talk can make a considerable impression upon farmers and build up confidence in the extension agent.

While not all agents will have the same natural ability to speak publicly, there is much that an agent can do to improve his performance.

Public Speaking

CHRISTOPHER M. WHITE

Public speaking is an important part of extension work

Most agents will be able to develop a good approach to public speaking with application and practice. The points listed below are a useful guide to public speaking for the extension agent.

Preparation

As with all extension activities, the agent must prepare himself beforehand for a public speaking engagement, no matter how unimportant the event may seem. This preparation includes checking the facts, figures and other information to be included in the presentation, organizing the material in a logical manner and preparing any supporting audio-visual material. It is also useful to check the place where the talk is to be given and to take into consideration the interests, needs and knowledge of the expected audience.

Some agents find it useful to rehearse the talk beforehand, or at least the main points. The agent should not try to memorize the whole talk; instead, he should write the main items in order for reference in large print on paper or on small cards. Another technique is to note the main points on an overhead projector sheet which can be uncovered step by step as the speech progresses. This serves a dual purpose: helping the agent in the presentation and providing a summary of the main topics for the audience.

Content

Great care should be taken with the content of a talk, both in terms of the words and expressions used and also in the logical sequence of what is to be said. The subject of the talk should be introduced and a general outline of the subject given. The main body of the talk should contain the key points that the agent wishes to make. He should not put too much content into a talk; a short, concise and well-thought-out talk will have far more effect than a lengthy, rambling presentation.

It is also important to take some care in the use of words and expressions and to adapt these, where possible, to the local context. Similarly, it is important to avoid the use of over-complex technical language or jargon, which might confuse the audience.

Delivery

This is the most critical part of public speaking and demands much care and attention. Confidence is very important in delivery. If an agent knows what he wants to say, prepares himself and says it clearly and effectively, he will probably give a good talk. It is useful to develop a friendly style and to talk to or with an audience, and not at them. Under no circumstances should an agent underestimate the intelligence of farmers and talk down to them.

During a talk, posture and body movement are important. Excessive fidgeting, gesticulating or other theatrical gestures can often distract listeners, although some gestures can be used effectively to emphasize a particular point. The agent should continually observe the faces of his audience and take note of signs of interest, boredom or disapproval. A talk should be a two-way communication process in that the agent should be sensitive to the effect he is producing and be prepared to react positively.

The agent should always ensure that his voice is loud and clear enough for all those present to hear. He should try not to be nervous, and should not apologize in advance for any shortcomings. A positive attitude to the delivery of a talk is very important.

Finally, the talk should be relatively short, about 15-20 minutes. Farmers will probably be unwilling to concentrate for longer than that, so the content should be restricted to the time available.

Questions and discussion

The audience should be told beforehand about a question and discussion period at the end of the talk, so it has time to prepare itself accordingly. The agent should encourage the audience to raise points, and be prepared to stimulate a discussion. A question and answer session alone should be avoided as this will discourage genuine dialogue and reduce the educational purpose of the talk. Moreover, the discussion should not drag on for too long, or be dominated by just a few farmers. It is also acceptable for the agent to direct questions to the audience in an effort to stimulate a two-way discussion.

Report writing

Extension agents are regularly called upon to write reports; indeed, a very common and serious constraint upon an extension agent's work is the frequency and number of reports which he will have to prepare. Invariably, these reports are demanded by an agent's superior and they can often take up a lot of time. Report writing is an important aspect of extension work and should be undertaken responsibly by the agent, but he should do all he can to avoid being overwhelmed by such reports or allowing the demands of report writing to interfere with practical extension work.

Like public speaking, report writing is a skill that the extension agent can develop and put to good use. As a guide to writing a report, the following are a few general hints to bear in mind.

Ensure that all the information and data which will go into the report are available and readily at hand.

Plan the report beforehand and decide upon its general content, format and style of presentation.

Structure the content in a logical order, introducing the purpose of the report, followed by the main substance, and then some concluding remarks.

Keep it brief! At whatever level the agent is operating, a brief, concise and well-structured report is far more useful and effective than a lengthy, rambling one.

Check over the report, once written, and ensure that the final version is clear, neat and easy to read.

An extension agent who can easily and quickly structure his report-writing duties will get through them with much less bother than the disorganized agent who prepares and writes his report as he goes along. Report writing can be time-consuming, and the extension agent will want to minimize the time spent on this bureaucratic responsibility.

The use of local leaders

A good extension agent will always try to enlist the support of local farmers in his extension work. In any extension organization, there will be only a small number of trained, professional extension agents within any one region, with responsibility for thousands of farming families. The solution is for extension agents to seek out and enlist the support of local people who have leadership qualities or influence within the area.

Local leaders can be of invaluable assistance to an extension agent in a number of ways. They can assume responsibility for certain activities in the agent's absence; help to organize local extension groups; assist directly in the spread of new ideas and practices by demostrating them in their fields; and generally serve as a point of contact between the agent and the farmers. By enlisting their help, the extension agent will have a chance to reach far more farmers than he could on his own. Working with local leaders also builds closer ties with local farmers, and encourages farmers' confidence in the extension service and their willingness to participate in extension activities.

Formal and informal local leaders

Extension agents will work locally with both formal and informal leaders. In Chapter 3, formal leaders were described as local people who hold some kind of formal position within the bureaucratic and administrative structure. Such leaders can include representatives or agents of government ministries; traditional chiefs or headmen; teachers; religious leaders; political officials; and officials of local institutions (e.g., cooperatives). The extension agent should try to interest such formal leaders in his activities and discuss appropriate parts of his work with them. If he can enlist their general support for his programme, then his extension activities in the area will have a firm basis. Certainly the extension agent should invite these formal leaders to extension meetings or other public extension activities.

When carrying out his extension programme and activities in the field, the extension agent will work with local informal leaders. Informal leaders are farmers, prominent in their area, who show the qualities and abilities which can be of use to an extension agent. Informal leaders will exist in a rural area, and by careful inquiry and observation, the extension agent should be able to identify them. Often, by talking to other farmers and asking whom they see as the natural leaders in the area, the agent will be able to identify the key farmers whose support will be invaluable in promoting extension activities.

Extension agents work closely with local leaders and involve them in extension activities

Selection of local leaders

Extension experience in different parts of the world has suggested the kinds of qualities and characteristics that the agent should look for in farmers who might be good local leaders. Clearly, an extension agent should not hastily invite a local farmer to become a leader of extension activities. He must give some thought to his choice. The following two lists are examples of the qualities to look for in local leaders.

1. Initiative to take the lead and give confidence to others.
   Intellect to understand issues and identify problems.
   Industry and energy, to work unselfishly with other farmers.
   Influence over others, and the ability to persuade and teach.
   Integrity and a sense of responsibility.

2. Experience in farming and modern agricultural practices.
   Educated and literate.
   Reliable and a regular attender of extension functions.
   Innovative and willing to try out new ideas.
   Trusted and liked by his fellow farmers.

The above lists are not meant as check-lists but do indicate the importance of selection, and the care that the agent must give to determining the qualities he feels are important. He must then seek out those farmers in his area who match the qualities.

Working with local leaders

The extension agent should take great care to develop the qualities mentioned above. His own relationship with local leaders will also be important and he should always try to be available to support and encourage their work. There are four main aspects of working with local leaders which the agent should keep in mind.

Inform local leaders of extension activities and proposals for new programmes, and keep them supplied with extension literature.

Visit them as often as is necessary — enough to ensure that they are not isolated or left on their own. Try to make the visits regular so that the leader can build them into his own work routine.

Train the local leaders in the aspects of extension activities with which they may be unfamiliar; formal training sessions can be set up at which the leaders will learn about a new practice, how to run a demonstration or how to hold a farmers' meeting.

Encourage local leaders to take the initiative and to begin to act with some independence. The more they can become recognized and effective, the better chance the extension agent will have of making an impact in the area.

An extension agent who has the use of the services of a group of good, efficient local leaders has a tremendous additional resource at his disposal and will be in a far better position to get extension work going in that area than if he had to work alone and unsupported.

Problems of working with local leaders

While working with local leaders can be of great benefit to the extension service, there are a number of potential problems the agent should be aware of. The agent should keep a watchful eye both on his own relationship with the local leaders and also on the performance of the leaders at field level. If the leaders are carefully selected and supervised, few difficulties will arise; yet it would be wise for the extension agent to keep a watch out for the following potential problems.

- If the agent spends too much time or concentrates a lot of effort on one or more of the local leaders, then the issue of favouritism may arise.

• The local leaders function as contact farmers who are expected to pass on the knowledge they have received from the agent. This flow of knowledge from the leader to the other farmers does not always work and the agent should pay particular attention to seeing that it does. If the leader is not functioning as a contact farmer, then the agent will need to investigate the reasons why.
• Some local leaders may become overconfident and domineering, and use their favoured position with the extension agent for their own individual gain.
• Some local leaders may be less capable than others and may make mistakes and give wrong advice to their fellow farmers. The agent should always ensure that a leader is well prepared before giving him responsibility for extension activities.

The above examples of working with local leaders are included, not to suggest that they will always occur but to remind the extension agent to keep a watchful eye on his use of local leaders and to be ready to respond to such problems if they do arise.

The extension agent is the key element in the whole extension process; without an agent in the field to guide, direct and supervise local extension activities, there would be no extension service available to farmers. The agent's role and relationship with the farmers are the critical aspects of this process and things cannot always be expected to go smoothly.

Extension experience in different parts of the world has stressed the agent's central importance and has highlighted a number of features which distinguish an effective extension agent from a less effective one. An effective agent:

• spends time in developing the skills and attributes of the farmers themselves, and does not merely concentrate on extension projects;
• gets out to visit and meet farmers and does not become an office bureaucrat;
• encourages local initiative and self-reliance and does not adopt a paternalistic attitude toward farmers;
• plans for the long term development of his area, and does not only seek quick results.

Again, the above are suggested as examples of issues that may arise in any extension area. In practice, most extension agents are committed people, working under difficult conditions, often with little support. The work of an extension agent demands the particular qualities of dedication, humility and hard work, and extension services should ensure that an agent is thoroughly prepared before he begins his extension activities.

7. The planning and evaluation of extension programmes

Extension programmes

In previous chapters, the methods and skills that an extension agent uses in his work with farmers and their families have been examined. It has been stressed that all extension activity requires careful planning if it is to be effective. No extension activity is planned in isolation; every demonstration, public meeting or film show is part of an overall extension programme through which an extension agent and farmers work toward the agricultural development of their area. In this chapter, a number of important principles that should guide the agent in planning and evaluating extension programmes will be considered.

An extension programme is a written statement which contains the following four elements:

Objectives which the agent expects to be achieved in the area within a specified period of time. This will often be a one-year period, to enable the agent to review the programme at the start of each farming year.
Means of achieving these objectives.
Resources that are needed to fulfil the programme.
Work plan indicating the schedule of extension activities that will lead to the fulfilment of the programme objectives.

An extension programme with clearly defined objectives is helpful to local farmers, the agent himself, his senior extension officers and other rural development agencies. For the farmers, it shows both what they can expect from the extension service and how effective the agent is. For the agent, the programme provides a firm basis for planning extension activities on a monthly and weekly basis and for anticipating well in advance what resources will be needed. Senior extension officers can use programmes to assess agents' performance, to offer advice for improvement and to justify requests for additional staff, equipment and funds. Furthermore, the programme helps other agencies to coordinate their activities with what the agent is doing. However, programmes can only be used in these ways if they are written and made available to all concerned.

A written programme is also useful when staff changes bring a new agent into the area. The new agent can use the programme to carry on from where his predecessor finished, thereby ensuring continuity of activities.

All organizations involved in agricultural development have their own procedures for planning, which can vary considerably. In particular, they can differ in the extent to which plans are made at national or local level. When considering the planning of extension programmes, two different forms can be distinguished.

Planning from below. Farmers, with their extension agents, make plans for developing local agriculture on the basis of local needs and potential, and then make requests for specific assistance from national and regional authorities.

Planning from above. The agent is simply expected to implement plans made at national level. He may, for example, be given a target number of hectares to be planted with improved seeds, or a specified number of farmers' groups to set up.

Successful extension programmes should include both planning approaches. National policies and programmes provide a framework within which the agent plans his local programmes, and they establish priorities, which he must follow. If a national priority is to increase production of arable food crops rather than livestock products, the agent will

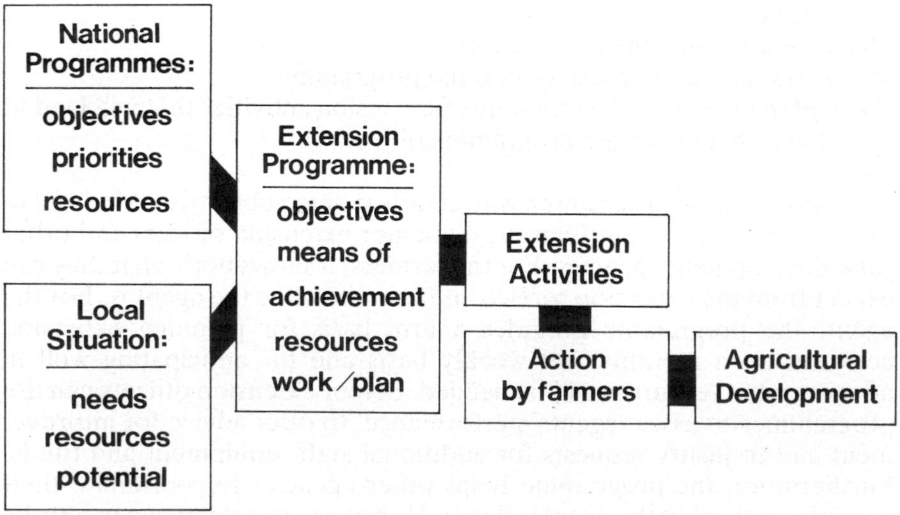

Extension programmes bring together national policies and local initiatives

give these crops a high priority in his own programme. National programmes will also make funds and inputs available for particular kinds of activity, which will influence the agent in his local planning decisions. But agricultural improvement comes from the willing action of farmers as they try to increase their own output and living standards. Local needs, therefore, provide the motivation for agricultural development, and must be taken into account in the planning of local extension programmes. Even in cases where the agent's freedom of decision is limited by national policy and directives, he must still prepare a programme that will enable him to fulfil these directives within his area.

In planning his extension programme, the agent should, therefore, balance national and local requirements. On the one hand, he should take note of national objectives but on the other hand, he should also work with local people so that the programme that emerges is theirs, and reflects their needs and what they want to see happen in the area and on their farms. This local involvement in planning is an important part of the educational process of extension. It stimulates a close analysis of farming problems and helps to build up motivation and self-confidence in using local resources to tackle them.

In some countries, agents work with formal, local-level committees when planning extension programmes. However this local involvement is achieved, the agent must take care that those who are involved can really represent the views and interests of all groups in the area. Committees often contain a high proportion of the more progressive, larger-scale farmers and are therefore inclined to promote programmes that fit the interests of these particular groups.

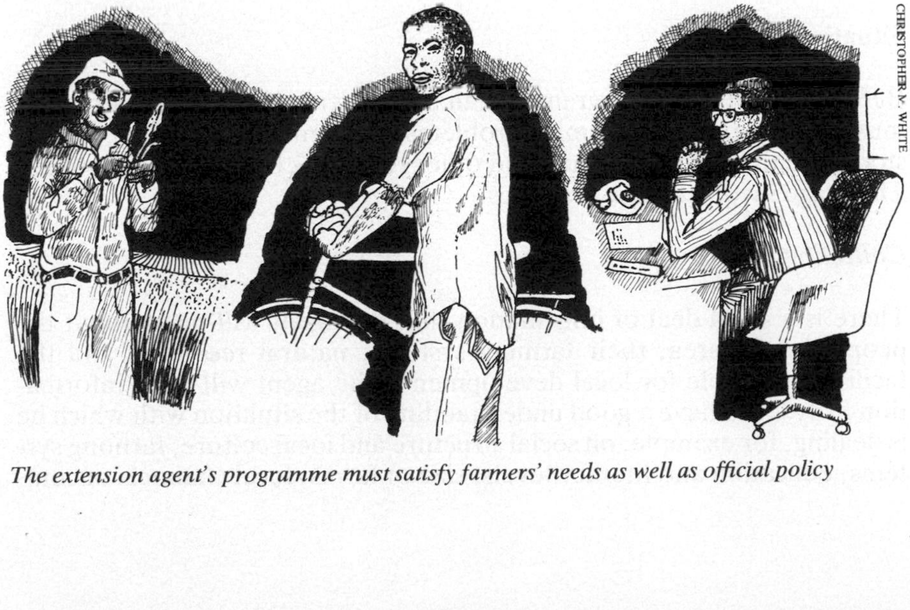

The extension agent's programme must satisfy farmers' needs as well as official policy

Stages in programme planning

Whatever particular procedures for programme planning are laid down by the extension organization, five distinct stages can be identified.

Analyse the present situation.
Set objectives for the extension programme.
Develop the programme by identifying what needs to be done to achieve the objectives, and then prepare a work plan.
Implement the programme by putting the work plan into effect.
Evaluate the programme and its achievements as a basis for planning future programmes.

This will then lead to a review of the situation and the planning of a new programme, which should build on the achievements and learn from the failures of the previous one.

The different stages of extension programme planning are interrelated and the planning does not always proceed neatly from one stage to another. Provisional objectives, for example, may be set during the situation analysis stage, but they may later be altered as new facts are collected and a deeper analysis leads to fuller understanding of the situation. Objectives may have to be altered still further as unexpected difficulties arise while the programme is being implemented. However, it is useful for the agent to think of programme planning as involving these five activities, each of which will be considered in more detail, as each can be broken down into smaller steps.

Situation analysis

Before an extension programme can be draw up, the existing situation must first be analysed. Farming problems and their causes must be understood and the natural, human and other resources of the area identified. This stage involves three activities.

Collecting facts

There is a good deal of information that the agent will need about the people in the area, their farming systems, natural resources and the facilities available for local development. The agent will need information in order to have a good understanding of the situation with which he is dealing, for example, on social structure and local culture, farming systems, education and literacy levels, size of farms, local channels of com-

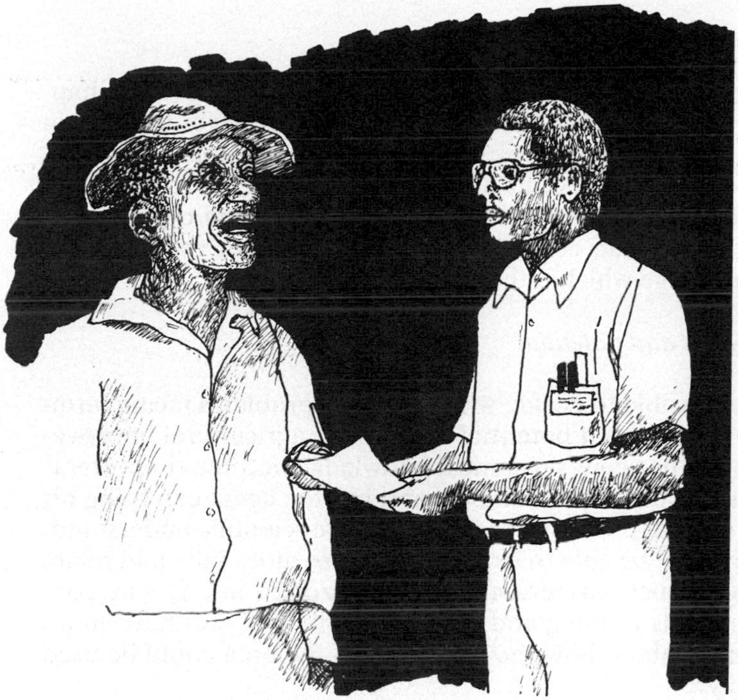

Collecting facts from the farmer

munication, transport facilities, local credit systems, marketing, health and nutrition levels, and crops and livestock.

These facts can be collected from a variety of sources. Reports of soil classification and land-use surveys, farm management studies, social surveys and previous programmes can provide a lot of useful background information. If the agent keeps good records of the farms in his area, he will have at his fingertips much of the information he needs. He can also obtain a lot of his information from farmers and local leaders. At public meetings, in group discussions and in contact with individual farmers, the agent should listen, ask questions and gradually build up a fuller understanding of the social, agricultural and economic features of the area.

Detailed social and economic surveys require complex questionnaires and statistical analysis, and are best left to professional researchers. But simple questionnaires can be used in this fact-finding exercise and it is helpful if the extension organization provides a standard list of questions or facts as a guide to the agent. One way in which farmers can be brought into the planning process at this early stage is for them to do some of the fact-finding themselves, using simple check-lists and questionnaires, for example, to catalogue natural resources in the area.

Analysing facts

Facts do not speak for themselves. It is necessary to ask why things happen in the way they do. If farmers report that yields have declined in recent years, the agent must look for other information that would suggest an explanation. Is it because of low rainfall, declining soil fertility, or poor seed? The agent must also separate fact from opinion and guesswork. He may obtain conflicting information from two different sources, and must judge which is the more reliable.

Identifying problems and potential

It should now be possible to decide what the main problems facing farmers in the area are, and what potential there is for agricultural improvement. This is where the agent's technical knowledge becomes important. Farmers may know what their problems are, but the agent can bring his own perception of local problems based on a more scientific understanding of farming. He will be able to explain problems more fully and relate them to processes which farmers may not realize are in any way connected. Because of his training and experience, he will also have more suggestions to make about how the resources of the area could be used more productively.

Throughout the situation analysis, the agent should avoid either, relying totally on his own expertise when interpreting facts and identifying problems, or leaving it entirely up to farmers to define local needs and possibilities for change. It should be a joint effort, with agents and farmers bringing their own experience and knowledge together to reach a full understanding. If farmers are not fully involved in these activities, the agent runs the risk of misinterpreting facts, wasting time in analysis and, almost certainly, of failing to gain the full support of farmers for the programme.

A full situation analysis is not needed every year. The basic facts about the area and the people will, in most situations, not change very much from year to year. However, the agent should review basic information each year and decide which parts of it need to be updated.

Setting objectives

Once the existing situation has been analysed, decisions can be made about the changes that should be brought about through an extension programme. The key questions are how will local problems be solved and how will local potential be developed. Solutions will require clear, realistic objectives which should be set in three stages.

Finding solutions

In looking for solutions to local problems, the agent should distinguish between technical solutions, involving improved inputs or simple changes in husbandry practice, and solutions which involve institutional changes, such as improved credit and marketing systems. Solutions involving institutional changes may require action by other agencies and at higher levels. While the agent should certainly suggest such solutions to those responsible, there may be little that can be done locally in isolation.

The sources of ideas for developing an area's potential include:

- the agent's own technical knowledge;
- farmers and agents from other areas who have tackled similar problems successfully;
- applied research which tests new ideas under farm conditions;
- national priorities and directives;
- projects which make funds available for particular activities.

F. MATTIOLI

Applied research is a valuable source of ideas for local extension programmes. These lettuce farmers in Hamala, Bahrain, are being trained in new cultivation practices with extension assistance

Selecting solutions

When selecting from among the range of solutions and possible improvements, agent and farmers should ensure that proposed solutions are:

Acceptable to farmers in the area.
Technically sound and tested by research and experience elsewhere.
Consistent with national policy, and with the local activities of other agencies.
Feasible within the time and with the resources available to farmers and the extension service.
Within the scope of the agent's ability and job description.

The agent may find that some problems will have no feasible or acceptable solution that can be implemented locally within the period of the extension programme. They may require legislation, action at other levels and by other agencies, or more research. The agent should lessen the effect of such problems where possible and act as a channel for putting forward the case for changes to those who have the power to make them.

Stating objectives

It should now be possible to state what the objectives of the extension programme are to be. But because his time and resources may be limited, the agent must decide which objectives have a higher priority than others. In doing so, he should consider national priorities and the size and distribution of the benefits that will arise from a given input of time and resources.

Wherever possible, objectives should be expressed in terms of amounts and numbers, rather than general statements. "Establish two groups of dairy farmers who will share new equipment and market their produce jointly", and "Increase the acreage of improved rice varieties from 60 to 120 acres" are more useful objectives than "Improve dairy farming techniques" or "Increase the use of improved rice seed". They give the agent firm targets to work toward, and a standard against which the effectiveness of the programme can be judged at the end of the year.

The objectives for an annual extension programme will state what should have been achieved by the end of the programme. These statements can be broken down into intermediate steps to be taken during the year in order to achieve the programme objectives. Again, the agent will have to make choices, selecting the most appropriate steps from several possibilities.

As the agent breaks down each programme objective into specific steps, he will in effect be preparing a schedule of extension activities for

DEVELOPING AN EXTENSION PROGRAMME

Problem

Local shortage of staple food.

Potential

Some local farmers have increased maize yields by 30-40 percent by using improved seed and animal manure; most farmers have plenty of manure but do not use it.

Solutions

A Increase size of small farms and introduce labour-saving techniques
— not feasible: no access to additional land.

B Encourage larger, mechanized farms to grow more food crops
— not acceptable: poorer farm families could not buy the food.

C Enable smaller farmers to increase yields
— feasible, using better varieties and tested husbandry improvements.

Preferred solution

C

Objective

Raise yields by 30 percent on 20 percent of the farms in the area in the first year.

Intermediate objectives

50 percent of farmers will learn of the benefits, and will acquire the skill, of using manure.

50 percent of farmers will learn the benefits of new varieties.

20 percent of farmers will plant improved varieties.

Plan of work

- Establish demonstration plots on ten farms.
- Hold public meetings and film shows in ten villages to show the benefits of new varieties and improved husbandry.
- Hold method demonstrations of manuring and correct spacing at the appropriate times.
- Visit farms during planting season.
- Hold result demonstrations on selected plots to encourage more farmers to try the new varieties and practices next year.

Support needed

- Subject-matter specialist to speak at public meetings, attend demonstrations and answer farmers' questions.
- Adequate supplies of seeds, at the right time, at local stores.
- Farm broadcasts to include relevant items at the appropriate time of year.

the programme period. He will decide what knowledge and skills the farmers will need; what additional technical information will be required from specialists and research workers; what extension methods should be used; and what resources and support he will need from his own and other agencies.

The simplified example on page 113 shows how the agent can develop an extension programme into a schedule of field-level activities.

When the planning is completed for other programme objectives, the agent can compile all the plans into an overall annual work plan. He may find that he cannot possibly do everything that all the individual plans require, so some of the lower priority objectives may have to be dropped, or scaled down. The annual work plan does not specify what the agent will be doing on each day during the year, but it should indicate when each extension activity will begin and end, and what resources will be needed for each.

Implementation

To implement the programme, the agent carries out the activities specified in the work plan. His detailed monthly and weekly plans will take account of progress and problems encountered in previous months. For example, the timing of some activities may have to be changed, or additional method demonstrations may be planned if more farmers than expected want to take part. An extension programme should be flexible enough to allow the agent to respond to circumstances in this way.

Evaluation

The agent will be constantly reviewing and evaluating his progress during the year. At the end of the year, a more thorough evaluation should be carried out in which the agent identifies how fully each objective has been achieved, and the reasons for any lack of progress. This evaluation, together with an up-dated situation analysis, provides the basis for planning the next year's programme.

Evaluating extension programmes

Evaluation is the process by which the effectiveness of extension is assessed. It is more than simply finding out what happened; it involves passing judgement on what happened. Was the outcome of the programme good enough? Was it better or worse than expected? Could more have been achieved?

Extension programmes are evaluated to (*a*) ascertain for the extension organization how well agents perform, so that their suitability for promotion may be assessed; (*b*) satisfy the government that public money spent on extension is being used effectively; and (*c*) permit the agent to learn from what has happened. Evaluation is a waste of time unless the results have an influence on future extension decisions.

One approach to evaluation is to ask if the programme's objectives were achieved. This is an important first step and one which is made easier if the programme had clear, precise objectives. If the answer is no, then there is no real basis on which to make improvements in future programmes. It is important, therefore, to ask why things turned out in the way they did. Only when that question is answered can the agent learn from the past. Agents should, therefore, ask questions about the following aspects of the programme.

Results. What happened as a result of the extension programme? Were they the results that were expected, and were there any unexpected results?

Inputs. Were all the planned inputs available and, if not, why?

Levels of evaluation

There are several levels of evaluation of extension programmes. At the most general level, the effect of extension on agricultural production, family incomes and standards of living can be evaluated. An increase in family living standards is usually an important ultimate goal of rural extension and it is, therefore, important to ask whether any increase has occurred. Evaluation of this kind involves measuring production and farm income for a representative sample of farm families, and then comparing the figures with previous levels. The changes revealed by these figures are then related to the extension inputs and activities during the programme.

However, extension is not the only factor that leads to higher production and living standards; changes in prices and in the availability of inputs are two of the many additional factors that affect the level of crop and of livestock production. Separating the effects of these various factors is a complex task and best left to specialist research and evaluation staff. Nevertheless, the agent should be aware of these economic changes and regularly ask himself how much his extension activities are contributing to the economic well-being of farmers and their families. He should

GOVERNMENT OF INDIA

Recording an evaluation discussion with farmers in India

also observe who is benefiting from extension. Is a broad cross-section of the farming population sharing the benefits, for example, or do one or two particular groups benefit most?

An intermediate level of evaluation is provided by the extension programme itself. Two questions are important here. Did the extension activities take place in the planned sequence and at the right time? Did these activities lead to the expected results? If the answers are negative, the agent should try to understand why. Perhaps he was over-ambitious about how many extension activities he could undertake, or maybe he did not receive the support he needed from other agencies. Whatever the reason, the agent will be able to learn from the evaluation process. He should be able to make more realistic plans in the future to ensure that the necessary support and inputs are provided.

Finally, the agent can evaluate at the level of each extension activity. All extension activities, such as demonstrations, talks or meetings with a farmers' group, have a purpose. The agent should try to check, wherever possible, not only how well the activity itself was conducted but whether the purpose was achieved. This will usually involve finding out whether the extension activity led to any changes in one or more of the following:

- awareness of a particular idea, possibility or problem;
- motivation to act in a particular way;
- knowledge about new farming practices;
- skills needed to adopt a new practice;
- behaviour by farmers and their families (such as new farming methods), or by an extension group (such as making an application for funds to carry out a group project or the preparation of a formal group constitution).

At all levels of evaluation, the agent needs to collect information to compare the situation after the activity with the situation existing before. He will already have assessed the situation before evaluation when deciding on the need for the extension activity. When planning a result demonstration, for example, he will have some idea of how many farmers in the area know about, are interested in, or have already adopted the particular practice that is to be demonstrated. However, he can obtain a more accurate assessment by asking those who attend the demonstration how much they already know and what experience they have had of the practice. By carrying out a similar assessment after the demonstration, he can collect the information he needs for evaluation.

Some effects can be assessed much sooner than others. Immediately after a public meeting, for example, the agent can talk to a few members of the audience and check how clearly they understood what he was saying. Changes in behaviour, on the other hand, will not happen at once and the agent must wait before checking these.

There are several ways of collecting information for evaluation at the village level.

Agent's reports. Whether or not a formal report of each extension activity is required from agents by their extension officers, the agent should make some notes on each activity for his own use, concentrating on his conduct of the activity and on points to note for future occasions.

Supervisors. It is not easy for an agent to assess how well he conducts an extension activity; in particular, he cannot see himself through the eyes of the farmers who attend. It is useful, therefore, to have constructive comments from a supervisor or colleague.

Discussions. Informal discussion with farmers after the extension activity will reveal their immediate reactions. It is often useful to record such discussions using a tape recorder for later transcription and fuller analysis.

Questionnaires. Simple check-lists and questionnaires can be used when the agent has the time and opportunity to carry out a more formal evalu-

The extension agent's records help to evaluate the programme

E. SCHWAB

ation of extension activities. Before a result demonstration on early planting, for example, the agent could prepare a list of four or five important facts that farmers should know after they have attended. By asking a sample of farmers questions on the facts, before and after the demonstration, the agent can assess its impact on farmers' knowledge.

Observation. Where changes in farming practice are concerned, observation is an accurate source of information. The agent can see whether or not his advice is being adopted on farms in the area.

Many extension organizations have their own formal procedures for evaluation. In some, the agent prepares a detailed plan of work each month on a standard form, showing what he plans to do each day during the month and how these activities fit in with his annual extension programme. The plan of work is then used as the basis for evaluation at the end of the month. Did he do all that he planned? Did he encounter any problems that he should take into account in the future? Is he on target in terms of progress toward his annual extension programme objectives? This procedure may be combined with a monthly meeting of agents in a particular district at which progress and problems in each area are discussed.

Whatever the formal procedures in a particular organization, however, the agent should think of evaluation as an attitude of mind. He should develop a readiness to ask what happened, why it happened and how it could be done better in the future. In this way, he will continue to learn and improve his extension work.

8. Extension and special target groups

Rural extension is concerned with the whole farming family, and extension programmes should cater for the needs and interests of the different members of the family. The present guide deals with extension in general and the principles and approaches it suggests should be relevant to all rural extension work. However, as was suggested in earlier chapters, different members of the family and of the community face their own particular obstacles and have their own special needs which should be taken into account in extension programmes.

In this final chapter, three special groups will be examined briefly — women, rural youth and the landless . The kinds of problems and issues which may arise with them both in agricultural and non-agricultural extension work will be identified. It should be stressed again that the basic content of the guide is relevant to extension work with all members of the farming community. It is felt, however, that these three groups do present particular problems for extension, which it would be useful to identify.

Extension and rural women

Agricultural extension services often relate more specifically to farmers (who are usually assumed to be men) and their various problems in the utilization and management of farm resources. Conversely, non-agricultural extension programmes are frequently directed toward women and seek to improve the use of resources within the home and family, or the care of the family's children. This common division, however, is not always appropriate. Many women are farmers in their own right, either because there is no man living with the family throughout the year or because women in some societies have their own land and their own crops for which they are responsible. Even where the head of the household is a man, women may do more than half the farm work.

In addition, therefore, to any agricultural extension programmes designed for rural women, it is important for agricultural extension to work with women, as well as men, to bring them the support, knowledge

Women are involved in agricultural production in Guatemala

and skills they need to improve their activities. In fact, over the past decade there has been an increasing concern to examine the role of women in rural development, to understand the particular contribution that women can and do make to this development, and to implement programmes and projects designed to improve women's lives. Until recently rural women have been neglected both in terms of our understanding of the particular kinds of problems that they face and also in extension action directed toward their problems. Most extension agents are men, and they perhaps lack a basic understanding of a woman's position in rural society. This position can be understood better by first considering the three basic roles of rural women.

- Economic, as producers of food and other goods for the family economy, and as a labour force for economic activities.
- Domestic, with responsibilities as wives and housekeepers to care for and manage the household economy.
- Reproductive, as mothers with responsibilities to reproduce family labour, care for children and look after their upbringing.

It is important for an extension agent to be aware of these three main roles which rural women have to assume, since they influence women's ability to participate in extension activities. They also indicate the kinds of extension support which would be of use. In rural areas, women often do a lot of work producing the family's main food crops. They are also responsible for storing and cooking the family's food, managing the domestic economy and supervising other economic activities, such as vegetable gardening or raising chickens, which are designed to increase the family's food resources. Moreover, they bear children and often have almost total responsibility for their care and upbringing. Rural women work very hard for long hours, usually for little reward, and are often neglected by extension services.

It is important for the extension agent to try to understand why there is often so little contact between his service and rural women. He should begin by analysing the situation and understanding the obstacles which prevent women becoming more involved in extension activities, and should take them into consideration when planning these activities. Recent studies on rural women have suggested a whole range of obstacles which rural women confront and which impede their greater involvement. These obstacles can be summarized as follows:

Cultural. Cultural obstacles are bound up in local customs and religious practice. In some societies, women are prohibited from conversing directly with non-family men. In others, custom forbids them to meet in public places, while in many, women are openly discouraged from participating in non-domestic activities.

Domestic. Domestic burdens are a severe handicap to women getting more involved in extension. Women have a full-time job contributing to the domestic economy and caring for and managing the family household.

Status. Women are generally accorded a lower status than men and are not encouraged or expected to play an active role in extension activities. Poor rural women find it almost impossible to break out from their ascribed status in order to have some voice in development.

The agent faces a difficult task in trying to incorporate women effectively into extension activities. But such is the importance of women's contribution to rural development that an extension service must work with women as much as possible. The agent must never consider rural women as inferior to men. They are not, but they do possess a different range of skills and abilities. Where possible, the agent should try to deal with both women and men at the same time. For example, he should encourage women to attend meetings and demonstrations; but women have different areas of responsibilty and the agent should direct his extension activities toward these responsibilities. It is imperative, however, that the agent study and understand the position of women in his extension area, and be sensitive to their particular needs and problems before embarking on any projects.

Although the agent should incorporate women as much as possible in his general extension activities, there will also be a need for projects formulated especially to support women's roles in rural society. Such projects could include:

Organization projects to build up and support local organizations representing women's interests and to encourage their activities, e.g., women's clubs or groups.
Production projects directly designed to assist those agricultural activities which are the responsibility of women, e.g., food crop production.
Health care projects to train the women and provide the facilities required for family health care, e.g., nutrition.
Income projects designed to help women to supplement their income, e.g., vegetable growing or handicrafts.

It is true that to date, men, as heads of families, have received the greater part of extension support, while women have benefited less and have been rarely encouraged to play an equal part in extension activities. But it is widely recognized that women do make a vital contribution to rural development and that extension should support this contribution. The real obstacles that women face must be understood, and extension agents, where possible, should seek out ways of channelling extension resources into activities which directly involve women.

Extension and rural youth

A large percentage of the population of many countries is under 18 years of age and the majority of the overall population lives in rural areas. It follows that young people make up a considerable portion of any rural population. These young people represent the farm families of the future, and it is essential that extension do something toward preparing them for that future. The alternative is for large numbers of young people to continue to drift away from rural areas and into the towns. For these reasons, extension agents should make a special effort to interest young people in their extension work; they should visit schools to talk about extension and should arrange tours of farm projects for young people to see what is going on.

Forestry project for young people in China

When talking of rural youth, both boys and girls are intended. Although there are no strict definitions, practice has shown that the boys and girls referred to are between 12 and 18 years of age. Many countries now have special extension efforts directed at these young people. Examples include the Village Polytechnics in Kenya, the Jamaican Youth Corps and the Youth Voluntary Service in India. The spirit of these programmes is to "catch them while they are still young", to show concern for the future of young people, and to try to get them started and established in the rural area.

Rural youth presents the extension worker with a number of particular problems. The agent must first try to understand these problems and how they affect young people's chances of self-improvement before embarking upon any youth-oriented extension activities. Extension experience with youth in different parts of the world has revealed the following kinds of common problems:

Motivation. As young people see the neglect and backwardness of the rural areas, they lose inspiration and often see little hope for their own futures.

Training. Many young people will have been formally educated but still lack the skills required to make a living in the rural areas. Often youth is ill-prepared and ill-equipped for the demands of rural development.

Involvement. Often young people feel isolated and unable to get involved in local rural development activities. They have no representation and no means of making their voice heard.

Opportunities. There are too few programmes which attempt to reach young people, or projects which seek to integrate the youth into rural development activities.

Obviously an agent will not be able to solve all these types of problems immediately but he can at least determine to give youth extension activities priority in his extension programme. He should make and maintain contact with the youth in his area, and he should quickly give some thought to developing stimulating extension activities for them. These activities often take the form of a club for young people, with specific projects particularly for their benefit.

Clubs for rural youth

Clubs have long been used by extension as a means to involve young people in rural development to propose projects for their benefit. The most widespread are the 4H Clubs (hands, health, head and heart) which began in the United States and which have now spread to many countries. In other countries similar clubs, such as the 4K in Kenya and the 4S in Panama, have been formed with the intention of bringing extension into closer contact with rural youth. These clubs are important as a first step to bring young people together; they give them an outlet to express their views and problems, and form a base from which to build for the future. Through them, young people also become used to working with the extension services, and they establish a relationship which will develop when they form their own farm families and need extension assistance.

When the extension agent considers forming a club for youth in his area, he should give the project a lot of thought and be mindful that these clubs have three main purposes.

Educational. In a variety of ways, both formal and informal, a club can be the means whereby young people can socialize and train for future life. Specific skills, leadership qualities and a general understanding of the problem of rural development can all be useful objectives of the club.

Economic. More particularly, a club can be used to instruct the youth in different aspects of agricultural practice, farm management or home economics. The club can also undertake specific projects designed to provide income for the youth.

Recreational. Not all the activities of the club should be serious. It should also encourage recreational activities and social events, e.g., sports, day-trips and even dances. In this way, the young people will enjoy the club, and will see it as an important part of their leisure time.

The agent should consult people locally before he forms a club, and ensure that it has the support of parents. He should also find a meeting-place and allocate some resources for its functioning. The agent's work with a club is very similar to his work with farmers' groups, and similar issues arise (see Chapter 5).

Two important aspects are the selection of a club leader and the internal organization of the club. Often the club leader will be a local, progressive farmer or even a teacher. The leader is not a member of the club and his task is to help to guide and support the club's activities. The leader (or leaders) will manage the club, help in the selection of projects and generally work in an advisory role to the club members. As the club develops, young people will need to be involved in its organization. This could be in the form of a committee of members, with a chairman, treasurer and secretary as committee officials. It is important for the agent to encourage the club to adopt an internal organization to provide some structure for discussion and for project work.

Projects with rural youth

Project work, either with the clubs or with rural youth in general, is the means by which young people can learn to do something instead of just listening to talks or lectures. The agent should encourage project activities with young people and allocate part of his local budget for such activities. These projects can be on an individual or a club basis, and should not be too ambitious initially. In terms of the approach to project work and to the steps involved in planning and implementation, the agent can consult other sections of this guide (see particularly Chapters 1 and 7).

A useful way of beginning project work is to take young people on a visit to a farmer, or to other agricultural projects, where they can see a particular activity for themselves. Local farmers are often most willing to collaborate with a group of keen club members. In addition, the agent could arrange for talks by other local people, or demonstrations to explain a project to them. It is important for the agent to be enthusiastic about project work and to try to involve youth in discussing and deciding what projects to undertake. Examples of youth projects that have been undertaken successfully in different parts of the world include poultry keeping, rabbit keeping, vegetable growing, handicrafts, fish-pond farming and home improvement.

Essentially, project work should be a learning experience. The projects are not only to provide useful additional income or food supplies, but should also be educational and a way of equipping young people with skills and knowledge useful for the future. It is important that the projects succeed, since failure could easily lead to early disillusionment. The extension agent, therefore, should give as careful attention to youth work as to more general extension work since he is really preparing and building for the future.

Finally, if an extension agent is to work with rural youth, he should have general sympathy with their views and ideals and feel at ease working with them. It might be better, therefore, for the younger agents in an extension service to take on the responsibilities of youth extension work. Young people will need to identify with the agent and be prepared to work with and trust him if extension work is to succeed.

Extension and the landless

In many areas where extension agents work, there will be farm families who are landless. The term landless includes not only people who have no land at all, but also families whose landholding is insufficient even for subsistence farming. Both these types of family are obliged to sell their family labour in order to make a living. It is not possible in this guide to give facts and figures on landlessness worldwide. The evidence is, however, that landlessness is quite common and increasing in many parts of the world, and it presents extension with a particular set of problems.

So far in this guide when extension and farmers have been discussed, and new technology, ideas and practices referred to, it has usually been assumed that farmers have access to the means (such as land) to take advantage of such innovations. Yet in many parts of the world a lot of farm families do not have direct access to the means of using the agricultural innovations suggested by the extension service. The question "What is extension's responsibility toward these families" must, therefore, be asked. These families present extension agents with an enormous challenge, which they can begin to confront if they try to understand the characteristics of landless families.

Extension and the landless: problems of contact and resources

- The landless lack an economic base on which to build any kind of future.
- They are dependent upon others for their livelihood, under conditions over which they have little control.
- The landless family's livelihood is precarious.
- They have little contact with extension or other government services.
- They have no influence over the decisions that affect their family's livelihood.
- Few organizations represent their interests.

It is not difficult to see that the above present a formidable challenge to extension agents who find landless families in the areas where they work. The agent will already have to implement policies and programmes; meet targets in order to achieve rural development objectives; give priority to increasing the productivity levels of existing resources; and work with the farmers who have the means to produce more. Nevertheless, where possible, extension services in general (and agents in particular) should give some thought to the plight of the landless. It will be a question of how much time the agent will have to devote, and how many resources he can allocate to efforts to improve the livelihood of the landless.

The agent should at least take time to study the problems of the landless in his area and be continually alert to ways in which he might make an impact on these problems. In several parts of the world — Nepal, Bangladesh and Peru — extension services have tried to tackle the problems of the landless. These efforts have been based on three main activities:

Organization. The extension agent encourages the formation of some kind of organization to represent the interests of the landless, and supports such organizations as he would support a farmers' group. Given the above characteristics of landless families, however, the agent must be prepared to devote a lot of time before any organization takes shape.

Resources. Where possible, the agent should try to make resources available to the landless to improve their smallholding or, if land becomes available, to help them to obtain the use of it. Small stock-raising projects will also provide some income, as will craft production projects with other members of the family.

Motivation. The landless often lack the will or the motivation to try to improve their circumstances. The extension agent can offer his support, show that he wants to give them some help, and generally try to encourage their interest in activities to improve their lives.

It would be wrong to suggest that an extension agent can solve the fundamental problems of the landless. The root causes of the situation of the landless lie in the rural society as a whole, and only when rural changes occur will the livelihood of the landless improve. The extension service and the agent must not turn away from a farm family merely because it does not have the resources to adopt new ideas immediately; extension is for the whole rural population. The agent must come to understand the structural obstacles which prevent the development of the landless, and he must try to tackle these obstacles whenever possible. Most of all, he must recognize the existence of the landless and offer extension's support in securing a more secure livelihood for them.

Bibliography

The preparation of this bibliography has been influenced by the fact that this guide is likely to be used by extension agents and their trainers who will have little time for extensive reading or research. It has been decided, therefore, not to produce a lengthy list of titles, most of which might, in any case, be difficult for the agent to obtain, but instead to suggest a small number of the more widely used texts on extension. The list has been divided into four broad types of extension literature.

Framework of rural development

COOMBS, P.H. & AHMED, M. *Attacking rural poverty.* World Bank, Johns Hopkins University Press. 1974

FOSTER, G. *Traditional cultures and the impact of technological change.* New York, Harper. 1962

LELE, U. *The design of rural development.* World Bank, Johns Hopkins University Press. 1979

LONG, N. *An introduction to the sociology of rural development.* London, Tavistock. 1977

WORLD BANK. *The assault on world poverty: problems of rural development.* Baltimore, Johns Hopkins University Press. 1975

Resource books on extension

CROUCH, B.R. & CHAMALA, S. *Extension education and rural development.* Chichester, W. Sussex, Wiley. 2 vols. 1981

JONES, G.E. & ROLLS, M.J. *Progress in rural extension and community development.* Chichester, W. Sussex, Wiley. 1981

MAUNDER, A.H. *Agricultural extension: a reference manual.* Rome, FAO. 1972

SAVILE, A.H. *Extension in rural communities.* Oxford University Press. 1965

Extension practice

BATTEN, T.R. *Non-directive approach to group and community work.* Oxford University Press. 1967

BENOR, D. & HARRISON, J.O. *Agricultural extension: the training and visit system.*
1977 Washington, World Bank.

DÍAZ BORDENAVE, J.E. *Communication and rural development.* Paris, Unesco.
1977

FAO. *Small farmers development manual.* Rome. 2 vols.
1979

FUGLESANG, A. *About understanding – ideas and observations on cross-cultural communica-*
1982 *tion.* Uppsala, Sweden, Dag Hammarskjold Foundation.

HAVELOCK, R.G. *Training for change agents.* Ann Arbor, University of Michigan, Institute
1973 of Social Research.

O'SULLIVAN-RYAN, J. & KAPLUN, M. (eds). *Communication methods to promote grass-*
1980 *roots participation.* Paris, Unesco.

STUART, M. & DUNN, A.M. Extension methods. In HAWKINS, M.S. *Agricultural livestock*
1982 *extension.* Canberra, Australian Universities International Development Programme, Vol. 2.

Journals

Agricultural information development bulletin
United Nations Economic and Social Commission for Asia and the Pacific (ESCAP), Bangkok, Thailand.

Bulletin
Agricultural Extension and Rural Development Centre, School of Education, University of Reading, UK.

Ceres
FAO, Rome.

Media in education and development
British Council Media Department, London, UK.

Rural extension, education and training abstracts
Commonwealth Agricultural Bureaux, Farnham Royal, Slough, UK.

Case-studies

The authors conclude this guide with a number of short case-studies of extension problems and practice. The case-studies are modified versions of a number of studies, which are kept at the Agricultural Extension and Rural Development Centre, School of Education, University of Reading, and are used in the authors' courses. The studies are drawn from different parts of the world and are based on actual situations.

In presenting these case-studies, the authors are not suggesting any models for extension work, or proposing any universally acceptable solutions. The studies are short, and present dilemmas and issues which an extension agent can expect to face during the course of his work. There are no correct answers to the case-studies and they should not be examined with that in mind. They are included in this guide to illustrate common extension situations, which can be critically examined by extension agents who use this text. The case-studies can serve also as a basis for discussion.

The authors believe that the case-studies touch upon many of the aspects and issues about extension that have been raised in this guide. For this reason, they believe that the studies will be a useful appendix and provide an opportunity to relate the content of the guide to actual extension work.

The authors would like to acknowledge material given to them by their colleagues Jancis Smithells and Ken Wilson-Jones in the preparation of these case-studies.

Case-study 1

The slogan that misfired

In a large cotton-growing area of the Near East, the Ministry of Agriculture has been facing problems of how to obtain maximum production. The growers have been following the official advice given (for timely sowing, spacing both within and between the rows, fertilizing and weeding) with good results for years, but they insist on sowing a handful of seeds in each hole instead of the five to ten recommended. In addition, they are unwilling to thin out the number of plants to the three-to-a-hole recommended, even though advice and demonstration on this matter have been going on almost as long as on the other matters. Since this practice reduces not only the yield but also the grade (and hence price and saleability) of the crop, the ministry has embarked on a massive publicity campaign including widespread use of posters and slogans.

One of the posters most favoured by the ministry, produced with the assistance of the local university's art school, clearly shows a large number of people eating from a very small communal bowl of food, all looking rather hungry. The slogan reads: "If too many try to eat from one bowl there is not enough for all—THIN YOUR CROP". Supplementary posters, to reinforce the message, showing, for example, better growth of plants in small clumps rather than in large ones, all carry the same slogan. Large numbers of posters were printed and distributed to agricultural offices, schools, village meeting-places and so on.

Although distributed in good time before the sowing season, the posters seem to have had no effect on farmers' actions. Worse, many of the recipients have not exhibited the posters, and other posters put up have been torn down or even defaced at night. What can be the trouble?

There could be a number of explanations.

1. The people have never been appealed to before by posters and do not understand their use or their message.
2. One of the faces depicted may resemble that of a respected notable.
3. The people are anti-government and see a political purpose in the poster.
4. There may be a deeper reason, which could be found out by investigation.

Fifty percent of the men and 15 percent of the women of the area are literate, and all the schoolchildren can read such a slogan. The people are of Islamic religion, relatively prosperous and not known to be anti-government to any abnormal extent.

Their cash crop is cotton, but they also grow more than their own requirements of grain and vegetables and keep a number of goats, sheep and cattle on the available grazing land. Most cropping is under irrigation. Crop residues and weeds are fed to animals.

Reasons come to mind why the cultivators may sow too heavily (e.g., if the soil forms a hard crust, a group of seedlings will break through more easily than a single one), but the resistance here is more than passive and based on more than technical objections.

Why do you think the local people have been so little influenced by the ministry's campaign, and what mistakes do you think were made? How might the extension service tackle the situation?

Case-study 2

An agent's dilemma

The holdings of most of the farmers in a newly established development area were too small and too badly fragmented to provide them with a decent livelihood. It was therefore decided to make available to them some additional land, on condition that they all agreed to give up their scattered plots for reallocation into larger individual, compact holdings.

The extension agent in that area had the job of introducing the plan, which he did at a meeting of the Development Area Council, on which the farmers were represented. He pointed out to the members of the council that if the farmers accepted the plan they would benefit in many ways: they would have more land; they would no longer have to waste time in going from one small plot to another; they would find a compact farm easier to drain and fence; and they could also make good use of tractors.

The agent then called a meeting of all the farmers. Although he took great pains to explain the plan's advantages very carefully, no one showed any enthusiasm for it. As a follow-up, therefore, he asked several of the most influential of the farmers to see him individually in his office in the hope that, if only he could convince them, they in turn would help to convince others but he failed with them also. They had, they said, farmed their existing lands for years and they intended to go on farming them. They knew exactly what they could produce. As for the new scheme, who knew what land he would get, or what kind of soil?

In the end, the agent had to abandon the scheme. Was it bound to fail, or might he have had a better chance of success if he had approached the problem differently?

Case-study 3

Credit: asset or liability

District X is one of above-average agricultural potential but the people have never shown great interest in production for market and seem to be happy with their traditional, mainly subsistence, cropping.

Their staple grain is maize, but yields are seldom anywhere near the potential, so that long periods of hunger are common in each rainy season immediately before the new harvest. Although this was hunger rather than actual famine, the authorities were concerned and in the early 1960s chemical fertilizers and improved varieties were introduced, with the object of increasing production. In order to make the change attractive, subsidies were offered to make the cost of the inputs acceptable to the farmer, and an encouraging number of farmers began to adopt the new inputs.

However, by the end of the 1960s, the country's economic position forced the dropping of the subsidies with the result that prices of inputs rose and their use dropped. Government feared that the situation would revert to what it had been before.

A new scheme was then introduced by which the Ministry of Agriculture was able to offer credit in what it defined as "low input areas". Under this scheme, fertilizer or seed would be sold to farmers and the cost, plus 15 percent interest, would be recovered at harvest time. The scheme was given publicity by extension staff, who welcomed it as a chance to gain the confidence of farmers who benefited from it.

Results were not encouraging. Some farmers used the inputs as intended but others did not use them at all. Others, it was discovered, sold the inputs for cash to farmers outside the "low input areas". Many were in default at the end of the season, and total production was far below the set target. A review of the system was ordered by the minister.

Why do you think the credit scheme failed? Could the extension staff have done anything to avoid failure?

Case-study 4

A problem of cattle

The people of the Sinkar tribe, occupying large lowland plains in Africa, are very independent and follow a nomadic way of life with their cattle. Their food is milk, or blood and milk, augmented by grain grown by the women at rainy-season camps. They occasionally eat meat at festivals. They are not very interested in cloth and they barter their cattle as exchange for wives and other needs. The result is that few cattle reach the market. Veterinary and other services are provided at government centres, and schools are also available to them. However, they do not change their tribal way of life and cattle numbers have increased alarmingly in recent years, so that pasture is seasonally scarce.

Background information

At the last census, tribal population was about 10 000 with 15 cattle per head, or upward of 150 000 cattle. This may be a serious underestimate as the tribe may be trying to evade cattle tax. The tribe's grazing area is about 483 × 80 km (300 × 100 miles), with seasonal availability of water, rather than grazing, as the limiting factor.

Veterinary services are available free for inoculation against certain endemic diseases. Other treatment has to be paid for but is subsidized. Education has to be paid for but children are lost from cattle-herding while at school. Health care is largely free but, like education, is only available at a few centres.

Veterinary effort, by removing the periodical epidemics which used to limit cattle numbers, has permitted a population explosion among the cattle, leading to serious over-grazing around the dry-season watering-points. Control of cattle numbers is now only maintained by seasonal starvation in times and years of drought, and the quality of beef, when it does reach the market, is poor.

There are a number of government cattle-buying points, and seasonal markets may be set up in the field. Prices for cattle are not high, because quality is low, but they are realistic. Cash is paid on the spot.

Of the educated tribesmen, some are in urban or other employment (including government services) while others have returned to rejoin the tribe and resume the nomadic life. They all form a lobby whenever the government tries to constrain tribal life or impose heavier cattle taxes.

The government has its own problems. The land is known to be capable of supporting a far higher population and of feeding the growing cities if farmed in a proper manner. Moreover, the country is in need of both food and export produce. From the politician's and city dweller's point of view, the Sinkar are monopolizing a valuable national resource merely to keep themselves in the manner to which they are accustomed, while others may be starving.

The problem is that the nation is short of meat and other foods. If the lowland plains were not occupied by the nomadic cattle-raising tribes, they could be exploited for cropping or as mixed farming areas.

You are sent as an extension agent to tackle this problem. How would you go about the task? What approach would you take and what kinds of difficulties do you think you might face?

Case-study 5

Introducing a cooperative

Republic X lies somewhere in Southeast Asia. Most of the land is flat and fairly densely settled, especially in the neighbourhood of the capital, and yields paddy and other food and export crops. Although not wealthy, the people are developing at a modest rate and appear to be generally happy.

However, there are also up-country hill areas, more thinly populated, inhabited by aboriginals of the country, whose language and culture are different from those of the rest of the country. Their areas have not been subjected to change and have in fact been rather neglected by government. Forty miles from the main lowland road and its trading centres, a secondary track leads to a village where about 80-90 families (about 600 people of all ages) live largely by subsistence farming. The village enjoys no regular services but there are two village shops, owned by merchants from outside, where they can sell their surplus produce and in return buy tools and other small wants (cloth, salt and non-essentials) for cash and seasonal credit.

An eminent overseas journalist has just visited the village and has written that the people have many complaints. They say that they are all in debt to the outside merchants, who overcharge them for their needs, charge heavy interest for credit, and make over 300 percent profit on the produce (mostly grain) which they buy from the farmers. Moreover, since there is no school, the children and adults are illiterate and unable to keep track of the merchants' records, so they fear that the merchants have been cheating as well.

If true, this report could have political repercussions, so a meeting is held at the provincial capital to consider what might be done in the village, which is typical of so much of the province. At this meeting, the provincial cooperative officer suggests that a cooperative society might be set up in the village on the following basis.

- A cooperative assistant, trained for such work, will be posted for six months, free of charge, to the village to set up the society.
- A local committee will be set up to manage the society.
- A contribution to the initial working capital of one year's per caput gross domestic product (currently about US $50) per registered member will be made by the government in order to get things moving.
- After the initial six months, the assistant will be withdrawn but periodic advice, assistance and audit supervision will be given by visiting members of the provincial cooperative office.

This suggestion seems sound to most of those attending the meeting, but the Ministry of Rural Development's representative feels that more facts are required before rushing in with a ready-made solution: a failure could be expensive not only in money but also in terms of popular confidence.

You are the extension agent who is sent to the village to investigate the truth of the report and recommend action to be taken. You may recommend acceptance or rejection of the cooperative as proposed, or any other solution from your own experience.

What do you need to find out and what questions will you need to ask?

Based on what you find out, what do you think your recommendation might be?

Case-study 6

The wells that failed

Viru is a rural community of about 2 000 people which lies in a fertile valley of the same name on the coast, 483 km (300 miles) north of Lima, the capital of Peru in South America. The Peruvian coast is bathed by the cold Humboldt current which, among other things, deprives the coastal region of rainfall and has created a narrow desert literally hundreds of kilometres long. However, intensive agriculture has been practised in Viru, and in many other villages in the same geographical situation, for thousands of years, thanks to a small river which flows from the Andes and enables irrigation to take place during the rainy season in the highlands (December to May). The water is often insufficient to irrigate all the fields. Yields are frequently low and crops sometimes fail altogether. With a more regular and abundant water supply, the farmers could harvest two crops a year rather than one.

Most of the farmers of Viru are too poor to undertake irrigation projects of their own, but through their political representatives they had been soliciting the Peruvian government for aid for many years. Many promises had been made but few had ever been fulfilled. Finally, the Government decided to drill six wells in strategic parts of the valley. The well-water was to be piped to the village for household needs and to implement a sewage system, and also to augment the supply of water for irrigation at those times when the river would be dry. A geological commission surveyed the area and selected the sites most likely to yield water. The first was on private land near the main irrigation ditch, but before operations could start it was necessary to repair and widen a road to haul the equipment to the drilling site. Although this was a community responsibility under the terms of the agreement and people were available, few offered to help. Indeed, the villagers hardly cooperated with the drilling team at all. Few villagers visited the site and general comments were highly critical of the whole operation. The technicians were somewhat surprised by such hostile attitudes and lack of interest, since the well was not costing the village anything. One well was drilled with considerable difficulty, but the project was abandoned in view of the lack of help and unfavourable response from the people of Viru.

Background information

In order to understand why this happened, it is necessary to explain a few facts about the culture of Viru and about the circumstances that existed at the time the well was being drilled.

The community was split between large and small landowners, and between natives of Viru and outsiders. The village was made up largely of small landowners and sharecroppers, mostly natives of the valley, who depended for their subsistence on small irrigated plots from 1 to 4 hectares in size; some of these farmers also sharecropped on large haciendas outside the village for commercial purposes, paying 25 percent of the crop as rent. The village also included a few large landowners, not all of whom were native-born, some of whom, through unscrupulous practices and unfair dealings, had managed to accumulate landholdings of considerable size. In some cases, these properties included parcels of land which formerly belonged to the community and the church, and which were secured through questionable deals.

National political circumstances had an important bearing on this case. The community was firmly split along political lines. The majority, the small farmers and sharecroppers, supported the governing political party (the liberals), one of whose main policies was to break up the large estates and return the lands to the people. However, the dominant power group, represented by the large landowners and the priest, was violently opposed to the governing party.

Municipal affairs were in the hands of a transitory board, a temporary body made up of outsiders and unqualified local people appointed by the government. The governing party was planning a bill to legalize municipal elections for the first time in Peruvian history and to do away with the old system of political appointees.

The value system of the villagers was such that most believed that natural phenomena, such as water supply from the rivers, were controlled by supernatural forces, as represented by the images of Catholic saints. If crops failed, people were being punished for their sins; if the harvest had been good, it was attributed to the saints, who controlled the weather, the insect pests and the water supply. Having been honoured by religious feasts, the saints were thus favourably disposed toward man.

On the basis of the information given above, why do you feel this well-drilling project failed?

If you had been an extension agent assigned to this scheme, would you have acted differently and, if so, what would you have done?

Index

Access to resources, 6, 8, 16, 20, 24, 128-30
Adoption of new practices, 33-37, 84
 barrier to, 37-40
 stages, 19-20
Advice
 acceptance of, 68, 83-84
 provision of, 11, 21, 26
Attitudes, 1, 17, 23, 24, 26, 35, 42
Audio cassettes, 51-52
Audio-visual aids 60-66, 80, 98
 advantages, 60
 guidelines for using, 65-66
 range, 60-65
 selection of, 65
Awareness, 19, 20, 45

Bahrain, 111
Bangladesh, 129
Blackboards, 62
Botswana, 33
Brazil, 75

Campaigns, 59
Cassava, 35
Ceremonies and festivals, 32-33, 40
Chalkboards, 62
Cinema vans, 53
Circular letters, 55
Clubs, 27, 71, 126-27
 leaders, 127
Cocoa, 34
Colombia, 75
Committees, 107, 127
Communication
 and cultural change, 35-36
 attracting attention, 55, 57-58
 between communities, 75
 between farmers, 88-89
 four elements of, 41-43
 importance of listening, 43-44, 71
 misunderstandings, 44-45
 planning, 41-43
 traditional means, 33, 59
 unintended, 43
Communities, 32
 conflict within, 25, 31
Community development, 16, 21
Community projects, 27
Confidence, 83-84
Contact farmers, 104
Credibility, 72
Credit, 15, 26, 109, 111
Cultural change, 33-37
Culture, 23, 29-33, 108, 123
 acquisition, 29-30
 respect for, 30, 33, 71

Decision-making, 11, 12, 13, 68, 92
Demonstration, 55, 71, 82-87, 113, 127
 basic principles, 85-87
 follow-up, 87
 method demonstration, 82-83
 planning, 85-86
 result demonstration, 83-84
 supervising, 86-87
Dependence, 5, 7, 78, 129
Development, 1-2
 (*see also* Rural development)
Dialogue, 18, 43, 48, 53, 100
Diffusion, 20, 104
Discussion, 51, 52, 54, 62, 80, 89, 99-100, 117
Displays, 58-59, 73

Egypt, 33
Evaluation, 87, 108, 114-19
 of extension activities, 114-19
 levels of, 115-17
 reasons for, 115
Exhibits, 58-59
Extension
 accountability, 13-14

Extension (*continued*)
  agricultural and non-agricultural, 21-22
  educational process, 9, 16-20, 67-68, 70, 85, 126, 128
  elements of, 10-13
  principles of, 13-16
  role in rural development, 7-8
  rural extension, 7-8, 20-22
  statements on, 9-10
  two-way process, 14-15
  working with people, 13
Extension agent
  as change agent, 92
  as educator, 70, 93, 128
  initiative, 96
  personal skills, 96
  qualities, 96-97
  relationship with farmers, 44, 67, 78, 91, 97, 103, 104, 127-28
  role of, 92-94, 112
  selection, 96-97
  training, 8, 94-95
  types of, 20-22, 104
Extension group
  membership, 77
  purpose, 77
  relationship with agent, 78
  size, 77
Extension methods, 67-90
  group methods, 75-89
  individual methods, 68-75
  selection of, 67-68
Extension programmes
  flexibility, 114
  identifying problems, 110
  objectives, 105, 108, 110-14
  planning, 62, 69-70, 95, 108-14
  priorities, 108-14, 129
  schedule of activities, 112, 116

Family, 26, 39
Facts, collection of, 108-9
Farm equipment, 33-34, 53-54, 82
Farm management, 11, 127
Farm practices, 30-31, 33-34
  as part of culture, 31
  changes in, 36-37
Farm visits, 68-72
  check-lists, 70, 71, 72
  follow-up, 71-72
  planning, 69-70
  purpose, 68-69
Farmers
  part-time, 37
  problems facing, 3, 5-6, 108-110
  progressive, 16, 107
  small-scale, 16, 26-27
  types of, 16, 20
Farmers' organizations, 11-12, 21
Farming system, 30-31, 36-37, 108
  changes in, 36-37
Field days, 87-88
Film, 52-54
  check-list for use, 54
  limitations of, 52-54
  16- and 8-mm, 52
Filmstrips, 65
Flannelgraphs, 64
Flip-charts, 63

Ghana, 34
Government policies, 13, 93, 95, 106-7
Group development, 21, 77-78, 130
Group extension methods, 75-89, 111
  advantages, 75-76
  group action, 76
  group learning, 75-76
Group meetings, 78-82
  check-list, 80
  conduct of, 81-82
  forms of, 79-80
  planning, 80
  purpose, 78-79
Groups in society, 27-28

Health and health services, 2, 7, 15, 21
Home economics, 122-24, 127

Incentives, 32
India, 125
Informal contacts, 74-75
Information
  collection of, 108-9, 117-18
  distortion of, 43
  for planning, 108
  on knowledge, attitudes and practices, 48, 57
  relevance of, 53, 57, 65
Information needs, 41, 48, 114
Inheritance, 32, 37
  of leadership, 28
Initiative, encouragement of, 12, 104
Innovations, 33-35, 82-84

Innovations (*continued*)
  awareness of, 19, 20, 45
Innovators, 20, 35
Institutional change, 111
Intercropping, 30

Jamaica, 125
Jargon, 45

Kenya, 125, 126
Kinship groups, 26
Knowledge
  creation of, 8
  transfer, 8, 11, 93

Land tenure, 32, 37
Landless families, 26, 128-29
  extension and, 128-30
Leaders
  formal, 28-29, 101
  informal, 29, 101
  local, 15, 28-29, 30-31, 101-4
  political, 15, 28
  selection of, 102-3
  working with, 40, 103-4
  youth club, 126-27
Leaflets, 43, 55
Learning, 16-20, 75-76, 85
Lectures, 80, 81, 97-100
Letters, 74

Magnetic boards, 64
Maize, 30, 39
Marketing, 15, 111
Mass media, 45-59
  advantages, 45-46
  characteristics, 45
  pre-testing, 48
  principles of, 46-48
  producers of, 48
  use in extension, 45
Meetings (*see* Group meetings)
Models, 61
Motivation, 12, 17, 117, 125, 130

Needs, 13-14, 17, 70, 107
Nepal, 30, 129
Newspapers, 56
Newsprint, 62
Nigeria, 30
Norms, 29
Notice-boards, 59, 73

Objects (as visual aids), 61-62
Office calls, 72-74
Office layout, 73
Overhead projector, 65, 98

Panama, 126
Papua New Guinea, 36
Participation, 7, 11, 59, 81-82, 85, 95
  in planning, 106-7, 110
Peru, 129
Pictures, misinterpretation of, 44, 45, 56-57
Population growth, 36-37
Posters, 43, 44-45, 55, 62
Prestige, 17
Pre-testing of visual aids, 63
Pride and dignity, 38
Printed media, 55-58
  pre-testing, 57
Public speaking, 97-100
  preparation, 98

Questionnaires, 117-18

Radio, 48-51
  check-lists for recording programmes, 50, 51
  limitations, 48-49
  local broadcasting, 49
  programme format, 50
  radio forums, 50
  use by the agent, 50
Record keeping, 71, 87, 109
Religion, 24-25, 28, 123
Reluctance to change, 37-40
Report writing, 100, 117
Research, 11, 15, 111
Rural development, 2-8, 15-16
  organizations, 15-16
  principles, 7-8
  problems, 5-6
  role of extension, 8
  strategies, 6-7

Schools, 16,125-26
Self-confidence, 13
Situation analysis, 108-10
Slides (as visual aids), 64-65
Skills
  learning new skills, 18-19, 82
  of farmers, 11, 16-17, 117
Social divisions, 24-27
Social expectations and obligations, 23, 29, 39
Social structure, 23-29, 108
South Africa, 33
Subject matter specialists, 113
Suspicion, 20, 35

Tape recorders, 50, 51-52
Target groups, 16, 77, 121-30
Taste of new crop varieties, 38-39
Telephone calls, 74
Television, 54
Timing
  of extension activities, 32-33, 37, 85, 114
  of farm operations, 31
Tours, 88-89
Tradition, respect for, 38, 40
Training of local leaders, 103
Training of rural youth, 126
Travel, 35
Trust, 72, 96

Updating technical information, 51
Urbanization, 37, 125

Values, 1, 38
Video, 55
Visual aids (*see* Audio-visual aids)

Whiteboards, 62
Women, 24, 121-24
  constraints faced by, 123
  extension and, 24, 121-24
  projects for, 124
  role and responsibilities, 24, 121-22

Youth, 15-16, 17, 124-28
  clubs, 126-27
  project, 127-28

Foto-tipo-lito Sagraf - Napoli